LEGO®
MINDSTORMS®
NXT

Mars Base Command

James Floyd Kelly
Christopher Smith

Apress®

LEGO® MINDSTORM® NXT: Mars Base Command

ISBN 978-1-4302-3804-1

ISBN 978-1-4302-3085-8 (eBook)

President and Publisher: Paul Manning
Lead Editor: Jonathan Gennick
Technical Reviewer: James Trobaugh
Editorial Board: Steve Anglin, Mark Beckner, Ewan Buckingham, Gary Cornell, Morgan Ertel, Jonathan Gennick, Jonathan Hassell, Robert Hutchinson, Michelle Lowman, James Markham, Matthew Moodie, Jeff Olson, Jeffrey Pepper, Douglas Pundick, Ben Renow-Clarke, Dominic Shakeshaft, Gwenan Spearing, Matt Wade, Tom Welsh
Coordinating Editor: Kelly Moritz
Copy Editor: Heather Lang
Compositor: Apress Production
Indexer: BIM Indexing & Proofreading Services
Artist: April Milne
Cover Designer: Anna Ishchenko

Distributed to the book trade worldwide by Springer Science+Business Media New York, 233 Spring Street, 6th Floor, New York, NY 10013. Phone 1-800-SPRINGER, fax (201) 348-4505, e-mail orders-ny@springer-sbm.com, or visit www.springeronline.com.

For information on translations, please e-mail rights@apress.com, or visit www.apress.com.

Apress and friends of ED books may be purchased in bulk for academic, corporate, or promotional use. eBook versions and licenses are also available for most titles. For more information, reference our Special Bulk Sales–eBook Licensing web page at www.apress.com/bulk-sales.

LEGO® MIND

For Decker and Sawyer – Life is fun, and you boys make mine even better.

–JFK

Contents at a Glance

Contents

About the Authors

 **James Floyd Kelly** is a writer from Atlanta, Georgia. He has degrees in English (University of West Florida, Pensacola, FL) and Industrial Engineering (Florida State University, Tallahassee, FL) and has found writing about technology to be the perfect job. James has written books on CNC machines and 3D printers, LEGO robotics, Open Source software, tablets, and much more. He also blogs for www.GeekDad.com and www.Makezine.com when he's not tinkering in his workshop. James is married to Ashley and has two great little boys who can't wait to get their hands on all the cool, shiny things in their dad's office.

 **Christopher R. Smith** enjoys exploring life's mysteries. His tenure as Senior Quality Assurance Inspector in the Shuttle Avionics Integration Laboratory at NASA's Johnson Space Center in Houston, Texas furnished opportunities where his innovations were recognized by NASA with a prestigious Space Act Award. Toiling over LEGO elements, CAD'ing building instructions, and working for LEGO as a MINDSTORMS Community Partner and Developer, provides the creative exploration he enjoys the most. Chris believes that everything is possible and our world is what we make of it.

About the Technical Reviewer

James J. Trobaugh has a degree in Computer Science and has been working as a software architect for 19 years. He lives in the Atlanta, Georgia area with his family.

He has been involved with FIRST LEGO League since 2004 as a coach for Team Super Awesome, and as a technical judge at LEGO World Festival. He is also the FLL director of the Forsyth Alliance in Forsyth County Georgia.

James started out as a LEGO hobbyist by founding the North Georgia LEGO Train Club in 1998 and has found that LEGO robotics is a natural blending of his LEGO hobby and his day job as a software architect. The added bonus is the joy of getting to share his love of technology not only with his own children but with kids in general.

Acknowledgments

I've written many books on LEGO Mindstorms robotics, and every one of them has its own list of individuals who deserve credit for the book you're holding in your hands.

My two most favorite people at Apress, Jonathan Gennick and Kelly Moritz, deserve a big round of applause. This was a tough book to write, and an even tougher book to finish. Both of them were great with the encouragement and even better when it came to dealing with the slow and difficult process of getting their author to finish up chapters. Their patience and understanding is greatly appreciated.

My technical editor, James Trobaugh, deserves another round of applause. James had to read over my chapters, build the models, test the challenges, and give me feedback on what worked, what didn't work, and what was just plain confusing. He's a true LEGO expert, in every sense, and the book would not have been possible without his involvement.

Next, I have Christopher Smith, my good friend from Texas, to thank for creating the great CAD images used in the building instruction chapters. I submitted photos to Chris of my models and crossed my fingers that he could make sense of my designs. He never failed to impress me, and the quality and look of the building instructions are top notch and beyond anything I could have hoped to do myself.

Finally, I want to thank my wife, Ashley. She has always supported my writing career, and she always manages to deal with her stressed out husband when deadlines loom (or pass).

Other folks have had a hand in making this book a reality, and you can find a complete list of their names and duties a few pages earlier in the book. Thanks to all of the Apress team for their roles in getting this book done and in your hands.

James Floyd Kelly

Introduction

Mars.

It's the next place that man will set foot on for the first time, hopefully around 2030. For now, a manned mission is still far away. But we will get there. And when we do, I feel certain that joining those astronauts for the ride will be a number of robotic companions.

Countless science fiction stories and movies take place on Mars, and the book you're holding is no exception. In this book, you'll find that Mars is a busy place, full of bases and people with jobs to do. It's a busy place, and it's a dangerous place, with robots taking on many hazardous duties and making life a little safer for humans.

But there are many things that are beyond the control of humans and robots, and these are the challenges that the members of Mars Base Command will face during their tour of duty on the red planet. Life is unpredictable on Earth, on the Moon, and most definitely on Mars. That's why Mars Base Command is looking for the best, brightest, and most creative individuals – those people who can think on their feet, make quick and logical decisions and, above all else, solve problems.

Are you up to the challenges that Mars will put before you? Are you willing to dive in, explore, and find the answers necessary to keep the bases on Mars functioning smoothly? Can you work together with your fellow specialists to examine, build, test, program, diagnose, and repair any challenge that comes your way?

Glad to hear it.

Welcome to Mars Base Command, Specialist. I've got a few special jobs in store for you.

Commander James Floyd Kelly
Mars Base Command

Who This Book Is For

This is a book of challenges. More specifically, these are challenges for a LEGO Mindstorms robot. The four challenges use parts found in a LEGO Educaiton Resource Set (Set # 9648). These challenges were designed using this Resource Set so as to avoid using parts from a LEGO Mindstorms robotics kit. Teams or individuals will build the various models used in the challenges and then build and program a robot to interact with the models.

While the book was written to address specific requests from teachers for challenges that would use the Resource Set and provide classroom activities, the challenges aren't just for the classroom. Individuals, after-school programs, and other organizations (such as scouting or home-school networks) will also find a number of activities here to engage kids of all ages with hours of hands-on opportunities.

How This Book Is Structured

The four challenges are broken up into small groupings of chapters.

The first chapter of each challenge is a fictional short story. This story introduces a problem that the Mars base personnel have encountered and provides clues or suggestions for how a robot might be used to solve the problem.

Additional chapters after each fiction story will provide the building instructions for the various models used to simulate the challenge encountered. Two to three chapters ared used to break up the building instructions into smaller, more manageable components. Readers will also find these chapters offer up suggestions for how to build and test robots to interact with the models.

The final chapter for each challenge will provide the rules and the suggested placement of the models for simulating the challenge. A point system is provided for scoring a competition and providing feedback to individuals or teams related to the success (or failure) of solving a particular challenge.

This pattern of short story, building instructions, rules and challenge setup continues throughout the book for four separate challenges.

Prerequisites

Experience with a LEGO Mindstorms robotics kit is definitely preferable, but these challenges can also be used as incentive to encourage students to dive deeper into the building and programming of LEGO robots.

Building and programming solutions for the challenges are not provided, so it's up to the individual or team members to increase their knowledge of the LEGO Mindstorms robotics kit in order to successfully complete the challenges.

CHAPTER 1

■ ■ ■

Plan B

We Fix It

Mars Base Alpha, Section D, Control Center
August 20, 2062 at 10:26 PM (Greenwich Mean Time)

"Would someone please turn off that alarm?" asked Lieutenant Raleigh. "I can't concentrate!"

The red flashing security lights on the ceiling were annoying, but it was the up-and-down wail of the alarm that was truly overwhelming. Base Alpha was on full alert, but only two of the five individuals seated in Section D were aware of the true extent of the danger facing the facility. Commander Evans was away for two days on an inspection of Mars Base Beta, which left Lieutenant Raleigh in command.

Sitting at her console, Kristie Raleigh was attempting to decipher the data scrolling down the flexi-screens. *These numbers just aren't making sense*, she thought. *According to these readings, the backup generators haven't kicked in yet.*

Four feet to Kristie's left, at another console, Engineering Specialist Brian Platt leaned over and quickly typed a command on his laser-projected virtual keyboard. The alarm in the room stopped, but Kristie could still hear it echoing from other parts of the base.

"Thank you," she said. "What's the status of the power consumption for Sections A through E?"

Emergency lighting in the room was running, a definite sign that the batteries were still operational.

But for how long? Kristie wondered. Twenty-five minutes earlier, a small meteor shower had brightened the sky. Meteor showers were nothing unusual, but today, a few larger fragments that didn't burn up entering the Martian atmosphere impacted with the power conduits connecting the fusion energy facility to Alpha. For the last fifteen minutes, her five-person team had struggled to reconnect the control systems that would give access to the fusion power station over 300 kilometers from Alpha.

With the base's primary power out, backup generators would normally kick in. But not today. Something was wrong, and the power supply figures were simply dwindling too fast.

"Power consumption is at 10 gigawatts per hour," replied Brian. He typed another command and leaned back in his chair, shaking his head. "We've got less than two hours of reserve battery power, Kristie."

Kristie paused and counted to five. It was an old stress-relief trick she learned in college, but back then, she would count to ten. Right now, she could only spare five seconds.

"OK, we've got to figure out why we're not generating backup power. Once we figure that out, we fix it. It's as simple as that," she replied.

"Video systems should still be operational, even on backup power, but they're not. I can only guess that those systems may have been damaged, too. There were numerous impacts between the fusion plant and here," said Brian.

"Someone's got to do a visual inspection," Kristie responded. "Who do we have over in PLS?"

Brian entered a quick command and scanned the screen. "The duty roster shows Davis and Rhodes as the response team for Power and Life Support."

"Get them on the radio," replied Kristie. "Let's just hope it's not as bad as the numbers are telling us."

It's Bad

Mars Base Alpha, Section E, Power and Life Support
August 20, 2062 at 10:42 PM (Greenwich Mean Time)

Kristie watched the clock on her console and tapped her fingers nervously. "Five minutes. They should have been able to get a visual for us by now," she said. "We've got to…"

The radio squawked, startling her. "Kristie, are you there?" It was the voice of Ian Davis, one of Alpha's power system engineers.

Kristie pressed the Send button on her console. "I'm here, Ian. Tell me something good," she said, releasing the button.

Another squawk issued from the speakers. "No can do, Kristie. It's bad. The computer systems regulating the backup generators took a direct hit. The systems are automated, so no one was in there, but it's going to take a day or more to get them back online."

Kristie heard Brian take a deep breath at the news. He rolled his chair closer to Kristie and pressed the Send button. "Ian, all we need are two, maybe three, generators back online—just enough to provide steady power for repairs. Any chance to get a few of them operational?"

"Sorry. The generators aren't designed to be run without the power redirect the systems provide. It was a bad design, and we'll fix it, but not today."

"Okay, Ian," replied Kristie. "Do what you can. And try to reduce power consumption for Section E. We'll get back to you. Out."

Kristie stood, closed her eyes, and took a deep breath.

"Any ideas?" asked Brian.

"Well, I had an old engineering professor who always said, 'Whenever possible, have a backup plan'," said Kristie.

Brian raised his eyebrows. "I thought our backup generators were our backup plan."

Kristie shook her head. "No. The generators are part of the standard emergency procedures when power is lost. We're going to have to deviate from those procedures a bit, I think."

"And that would involve…?" asked Brian.

"We go to Plan B," replied Kristie. "Let's move."

Will That Thing Even Work?

Mars Base Alpha, Section B, Storage and External Access
August 20, 2062 at 10:58 PM (Greenwich Mean Time)

"You have got to be kidding," said Brian. "That thing hasn't been used in over five years."

Kristie pulled back the tarp and threw it behind her. "I started out in Environmental Maintenance and Control. I learned how to program on a model just like this one. It's got voice control and level-five sensor capabilities and can withstand the temperature out there," she replied.

"But will that thing even work?" asked Brian. "And even if it does, what's the plan?"

Kristie reconnected the internal batteries and flipped the power switch. A small whine was heard from within as processors and cooling fans began to boot up, and within a few seconds, external system lights were all green.

Kristie looked at Brian and smiled. "So far, so good," she said. "Let's try a few commands. E-M-3, status report please."

The large robot gave a small lurch to the side. "Environmental Maintenance Bot 3 ready," the robot responded in a pleasant voice that reminded Kristie of her old archaeology professor, Dr. Hicks.

"OK, it works," said Brian. "Now what? If Ian can't fix the generators, this robot certainly isn't going to do it."

"Forget about the generators for now. The situation has changed," said Kristie. She dropped to one knee and looked directly into the robot's visual processor. "E-M-3, I've got a job for you."

Now, We Wait

Mars Base Alpha, Section D, Control Center
August 20, 2062 at 11:18 PM (Greenwich mean time)

From the safety of the control center, Kristie and Brian watched E-M-3 through the small window ports as it exited the cargo bay and disappeared around the corner of Section E. Because video systems were still down, they would be unable to verify all of the robot's movements. Kristie had done the best job she could in programming the robot with the details of its destination and the job it needed to perform, but there were always going to be unknown elements. Kristie just hoped that the robot would be able to manage.

As E-M-3 rolled across the Mars landscape, Kristie spoke into the microphone on a nearby console. "E-M-3, can you hear me?"

"Affirmative," the robot responded over the speakers.

"Good. When you've reached the junction at Collector A-1, stop and report in," said Kristie.

"Affirmative."

"How long will it take to reach the first collector?" asked Brian. "We've only got about 70 minutes or so of backup power."

"By my estimates, it'll take about ten minutes. But each collector after that will only take a minute or two. How many do we need rerouted to gain a stable power level?" asked Kristie.

Brian did some quick calculations on a terminal. "Well, all of the sections have reduced their power consumption like we asked," he responded. "If we can get six of the collectors feeding power to the base systems, we should be OK. But, since the robot is out there, we might as well have it bring all 40 collectors online."

"Agreed. Now, we wait."

Kristie sat down in her chair and rocked nervously. Four years ago, the base had rerouted power from the solar cell farms. Solar Collectors A-1 through A-10 were now feeding power to five remote monitoring stations a few kilometers from Alpha, and the 20 collectors from Farm B and Farm C were providing extra power to the heaters at the hydroponics facility just to the south of the base. Farm D, with its ten collectors, was dedicated to providing power to the garage's seven rovers. *No one will be going for a drive today*, thought Kristie.

The plan was simple, but they were relying on a robot to complete it. The danger was simply too high for a human to attempt the re-routing of the collectors' power.

Each solar collector was matched to a power interrupt device and a power redirect. First, E-M-3 would need to take a collector offline by shutting down its power interrupt device. But each PID maintained a large electrical charge that was slow to dissipate. If E-M-3 made contact with the top of the charged PID, the robot's circuits would be damaged beyond repair.

Once the PID was offline, E-M-3 would then proceed to the power redirect, a simple switch for rerouting the power to the base.

Next, E-M-3 would manually rotate the collector back to its original bearing; a drawback to taking the PID offline was that the collector would rotate away from the Sun to protect the solar cells when not collecting energy.

After reorienting the collector, E-M-3 would then reengage the PID for power flow. If everything was done correctly, E-M-3's actions would redirect the energy gathered by the collector to Mars Base Alpha. Of course, E-M-3 would need to successfully redirect power from five more collectors.

"E-M-3 status: Arrival at Collector A-1 confirmed," came the voice from the speakers.

Proceed

Mars Base Alpha, Section D, Control Center
August 20, 2062 at11:29 PM (Greenwich mean time)

Kristie rolled her chair over to the communications console and spoke into the microphone. "E-M-3, confirm your exact position."

"E-M-3 status: North facing at coordinates 19.40 degrees north, 33.12 degrees west," replied the robot.

Brian consulted the map on the screen in front of him. The robot was indeed at the entrance to the power conduit switch for the collector. He nodded to Kristie. "In position."

"Okay, E-M-3. Please confirm your objectives," Kristie said.

"E-M-3 objectives: Disengage PID. Configure power redirect for Alpha supply. Reorient collector to optimum bearing. Reengage PID," replied E-M-3.

"And after completing Collector A-1, where will you proceed, E-M-3?" asked Kristie.

"E-M-3 objective: Proceed with identical process on Collectors A-2 through A-10, Collectors B-1 through B-10, Collectors C-1 through C-10, Collectors D-1 through D-10," replied E-M-3.

Kristie nodded at Brian. "Does that sound right to you?" she asked.

"Sure does. Let's get moving; six collectors need to be rerouted in 58 minutes."

Kristie nodded and spoke into the microphone. "E-M-3, proceed with your objectives."

"Affirmative."

And the radio went silent.

Operational

Mars Base Alpha, Solar Farm A, Collector A-1
August 20, 2062 at 11:32 PM (Greenwich Mean Time)

```
> System Status: Idle. . .
> Initializing Program Collector1. . .
> Run Program Collector1. . .
> System Status: Operational. . .
```

Environmental Maintenance Bot 3 did not hesitate in its duties, and the robot began moving into the electrical conduit cage for Collector A-1.

Your Turn

Mars Base Alpha, Solar Farm A, Collector A-1
August 20, 2062

Was E-M-3 successful? It's time to find out. You're in charge now, and the fate of Mars Base Alpha is in your hands. Continue on with Chapters 2, 3, and 4 to build the simulation environment for your own Environmental Maintenance Bot. You'll build a power interrupt device, a power redirect, and a solar collector. After building these three devices, you'll next need to construct and program a robot to successfully complete a set of missions that are described in Chapter 5.

Good luck!

CHAPTER 2

■ ■ ■

Creating the Power Interrupt Device

To successfully complete the Mars Base Alpha challenge, you will be required to construct a robot that will interact with three models. These models are built using parts from the LEGO Education Resource Set.

The first model you must prepare is called the power interrupt device (PID), and you will find complete building instructions for the PID in this chapter. Figure 2-1 shows what the completed PID will look like.

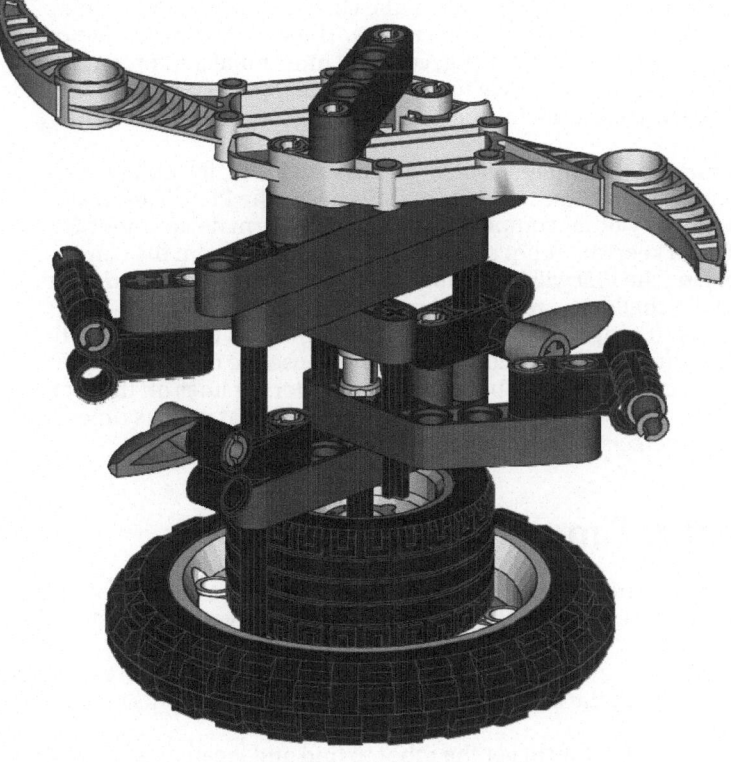

Figure 2-1. *The power interrupt device (PID) fully assembled*

The PID has a few moving parts. On top, you'll find two spinning claws that your robot must avoid at all costs. Below the claws is a swiveling mechanism; when the PID is engaged, the ends of the swiveling arms are touching the two orange pieces. When it's disengaged, the ends of the swiveling arms are not touching the two orange pieces.

Chapter 5 will provide details on the placement and operation of the PID in the challenge. You will also find information related to the challenge area; this can be quickly put together with nothing but the three assembled models for the Mars Base Alpha challenge and some tape to define the boundaries. You will also find details on a mat that can be downloaded (for free) and printed out in color or black and white and used as the challenge area.

Tackling the PID Challenge

After assembling the power interrupt device, you'll want to spend some time examining it. I suggest that you place it on a solid surface and take some measurements that will help you when it comes time to build a robot to solve the challenge.

Notice that the PID's most dangerous part is at the very top—a set of spinning claws (blades) that would easily damage a real robot if they were real. You'll want to make certain that any part of your robot that approaches the PID does not touch the blades, as the challenge rules will either disqualify the robot or result in penalty points (depending on the level of complexity of the challenge you wish to run).

Below the blades is the part of the model that will award you points. The swiveling arms start out touching the small orange pieces, indicating that the PID is engaged. You must build and program a robot that can safely disengage the PID by moving the swiveling arm in such a way that the ends are not touching the orange pieces. And remember, your robot must do this while not touching the spinning claws on top.

Keep in mind that the swiveling arms are extremely sensitive and easy to move; it won't take much for your robot to move them away from the orange pieces (and thus disengage the PID). This means you are not likely to need the use of a motor or any other complex device. A simple arm device mounted on the robot can easily disengage the PID and keep the robot's main body safely away from the claws.

The most difficult tasks in disengaging the PID will be having your robot identify it and orient itself in such a way that it does not go out of the challenge area boundaries. How can you do this?

It will involve taking measurements and programming your robot to move and turn properly to place itself where it can safely disengage the PID. If you choose to print and use the mat provided in Chapter 5, you could provide some visual cues for your robot, such as the colors or lines on the mat. Using a Light or Color sensor, you can program your robot to start, stop, and turn at various times depending on lines and colors the sensor detects on the mat or ground.

Taking One Challenge at a Time

After you've successfully built the power interrupt device, you may consider building a small robot or using the standard tribot (instructions are included with the NXT-G software) to try to solve just the PID challenge.

Rather than building all three models used in the Mars Base Alpha challenge, consider breaking the challenge into parts and attempting to work out the design and programming requirements needed to interact with only one model at a time.

In this instance, your goal would be to find a way to get the robot to find and locate the PID in the challenge area and then disengage the device. Once you have a robot that can successfully complete that challenge, move on to the next model, the power redirect (PR) covered in Chapter 3.

Don't get frustrated. Keep in mind the single goal here—move the swiveling arms in such a way that they are not touching the orange pieces. Don't overcomplicate your robot design here. Just get the robot from the starting position (see Chapter 5) to a suitable location on the mat to interact with the PID. Do that one task correctly, and you'll be able to do it again no matter the final design of your challenge robot.

Building the Power Interrupt Device

In this section, you will find the building instructions for the power interrupt device. Each image shows the pieces you will need to locate in the LEGO Education Resource Set and their quantities. You will also see where these pieces are to be placed.

If you are uncertain about the placement of a piece in a figure, jump ahead to the next image or even go back to a previous image to make certain you've built the model correctly so far. Examine the figures carefully, and you should be able to correctly identify the pieces and their final location on the model.

0

1

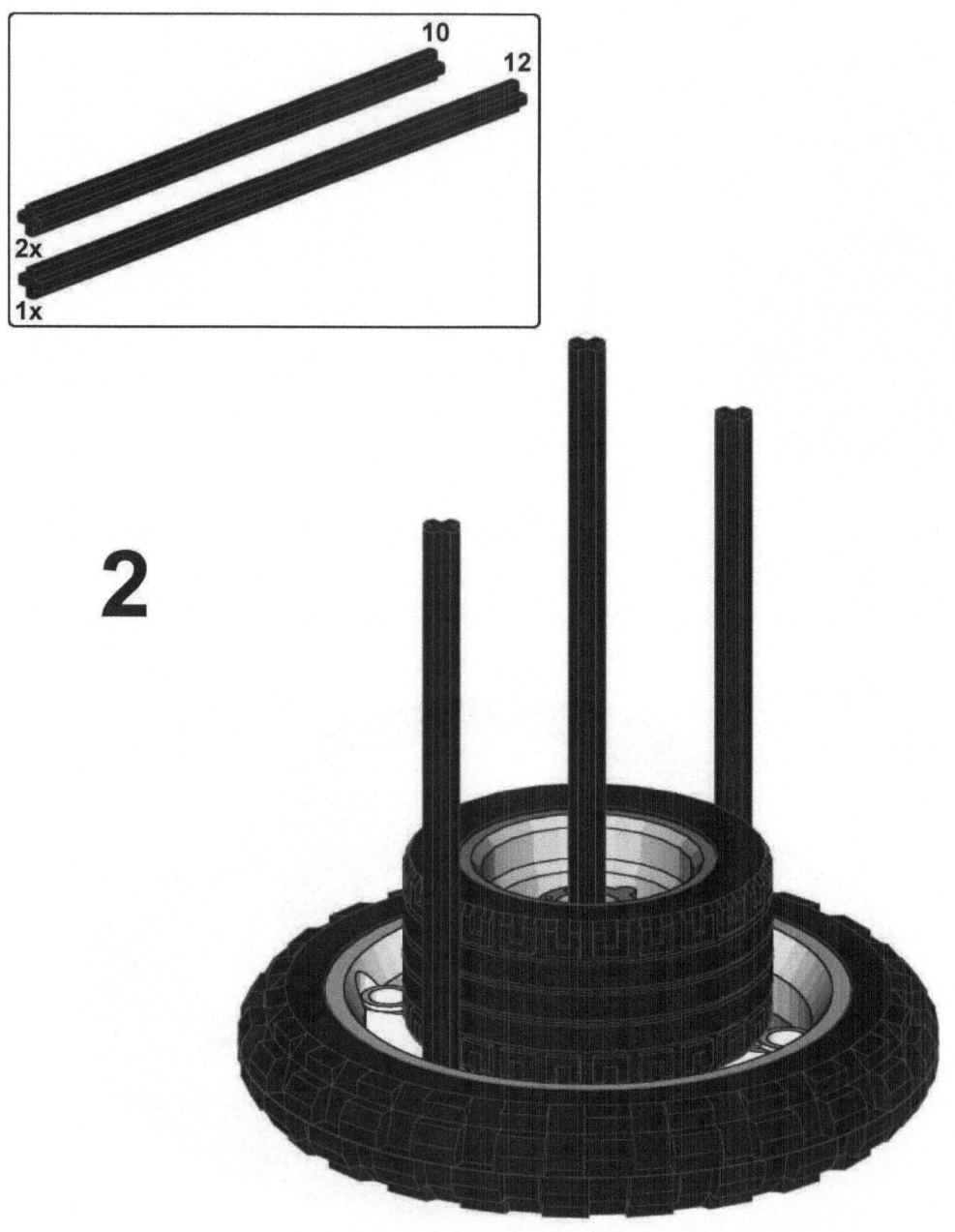

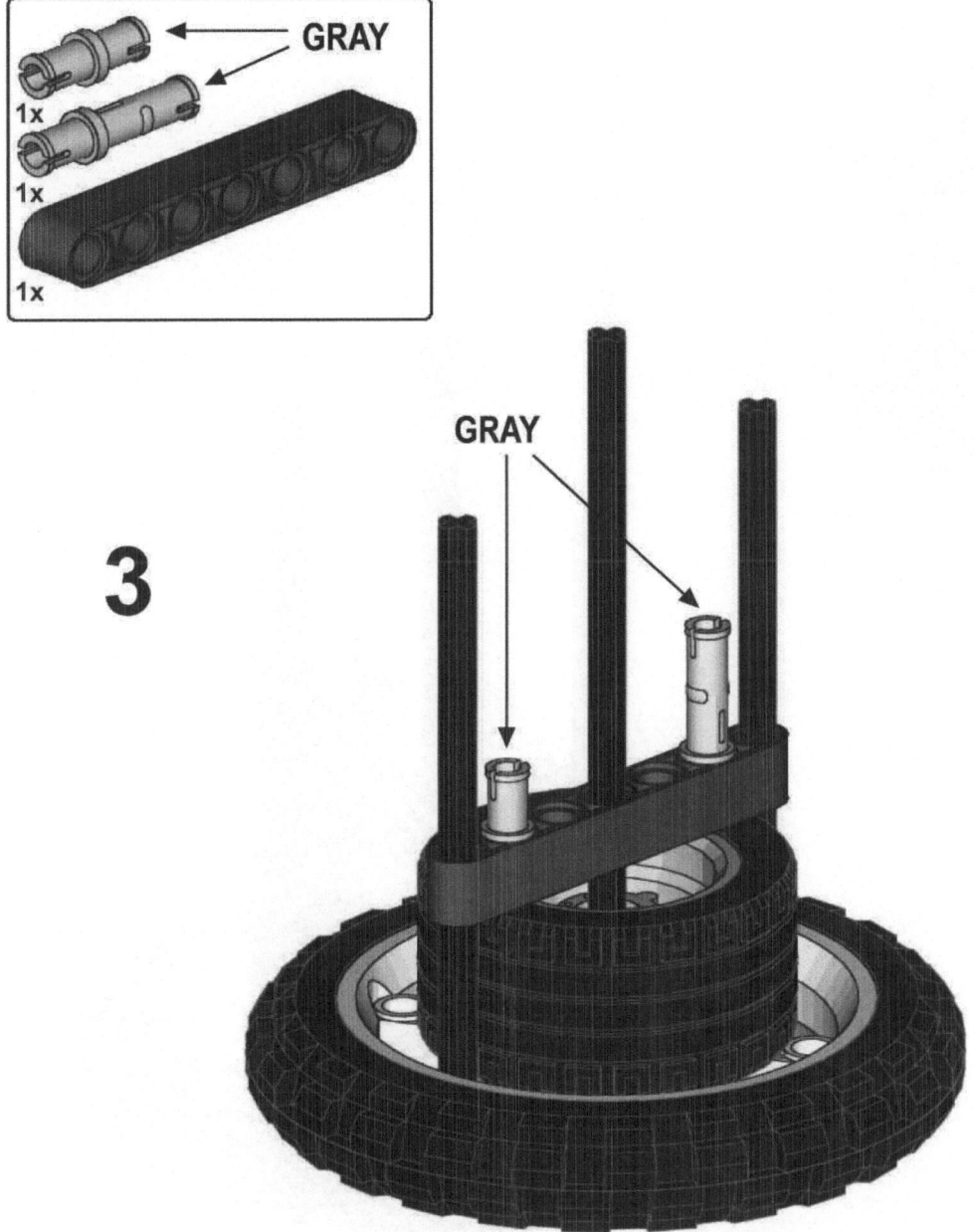

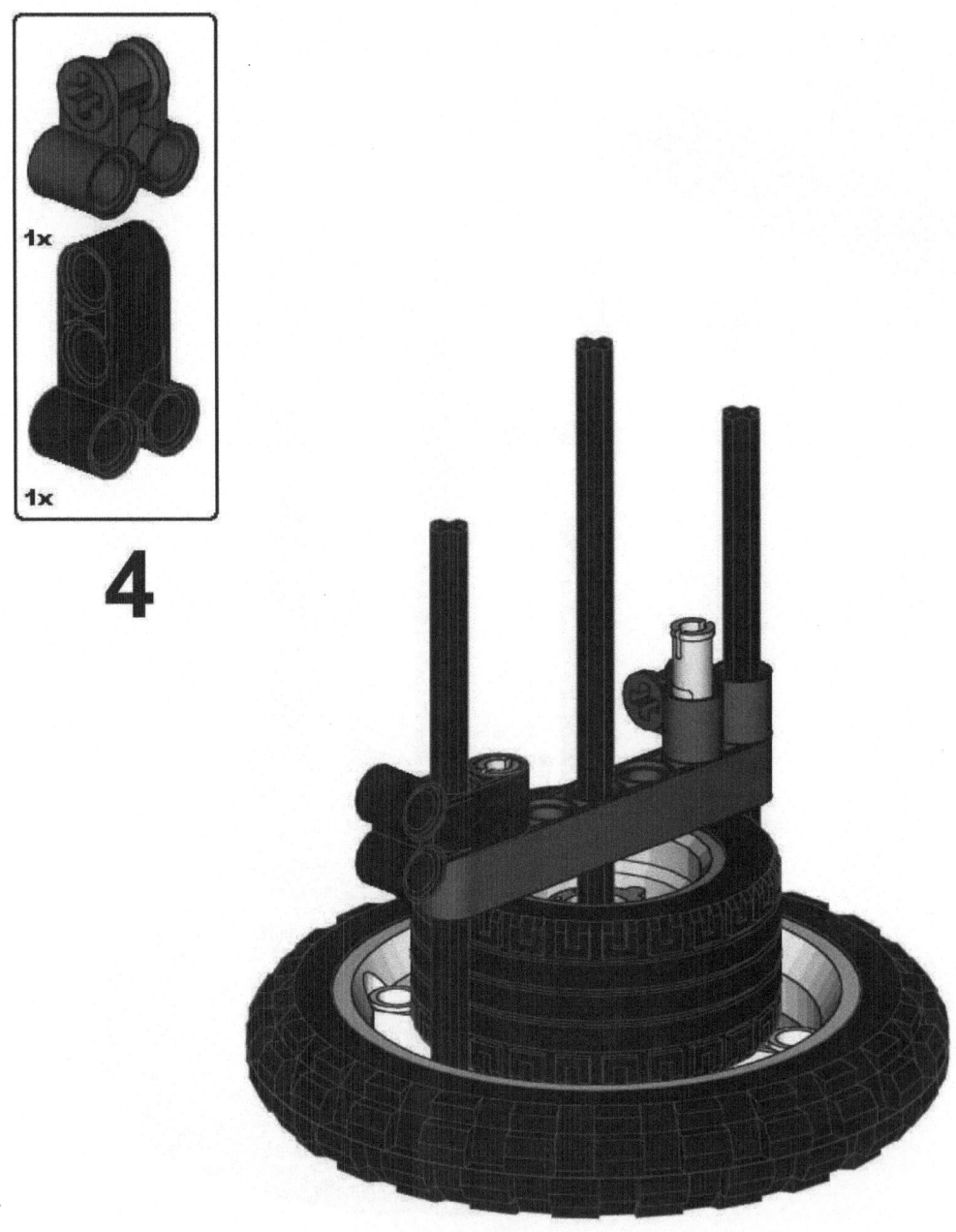

4

1x

5

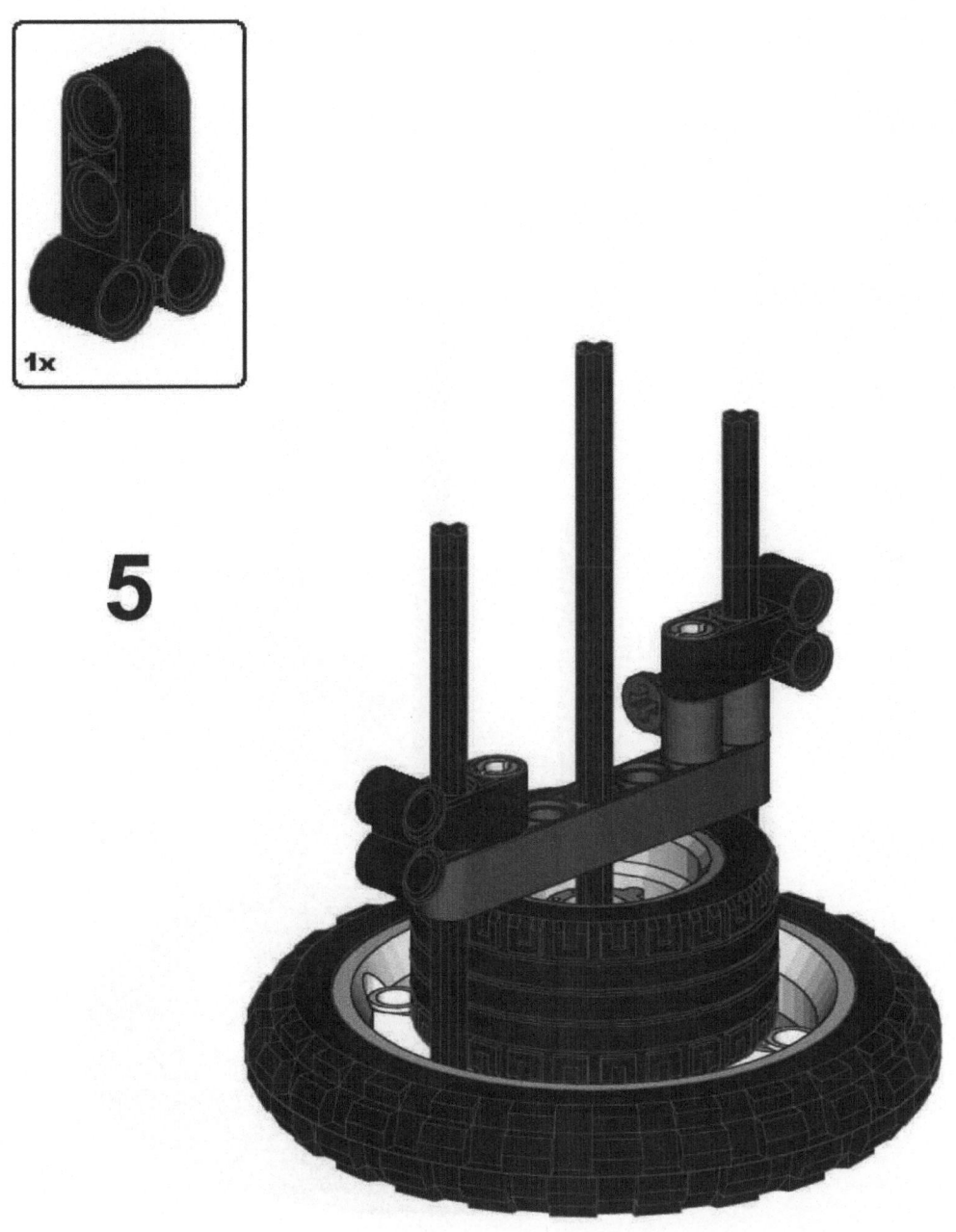

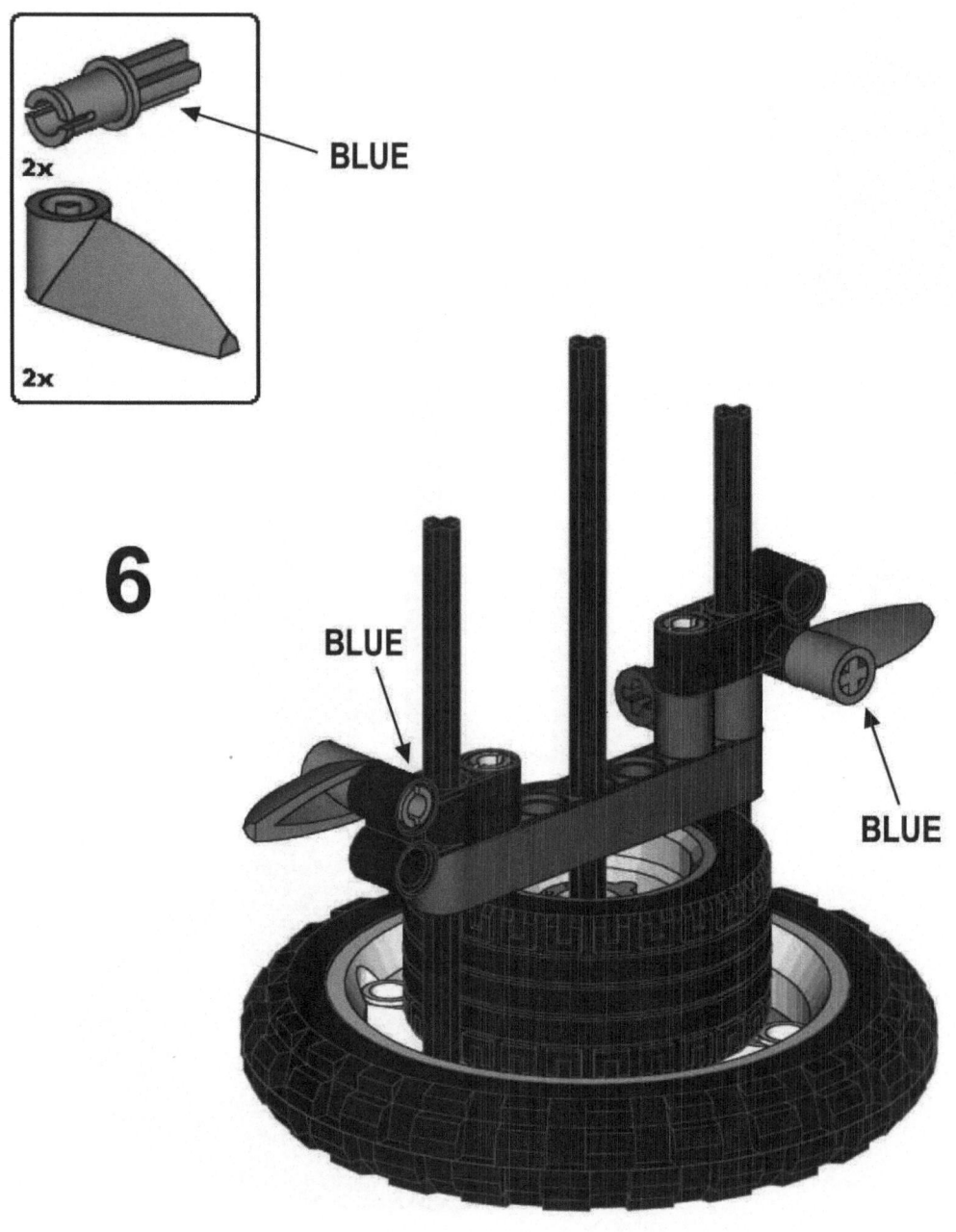

BLUE

2x

2x

6

BLUE

BLUE

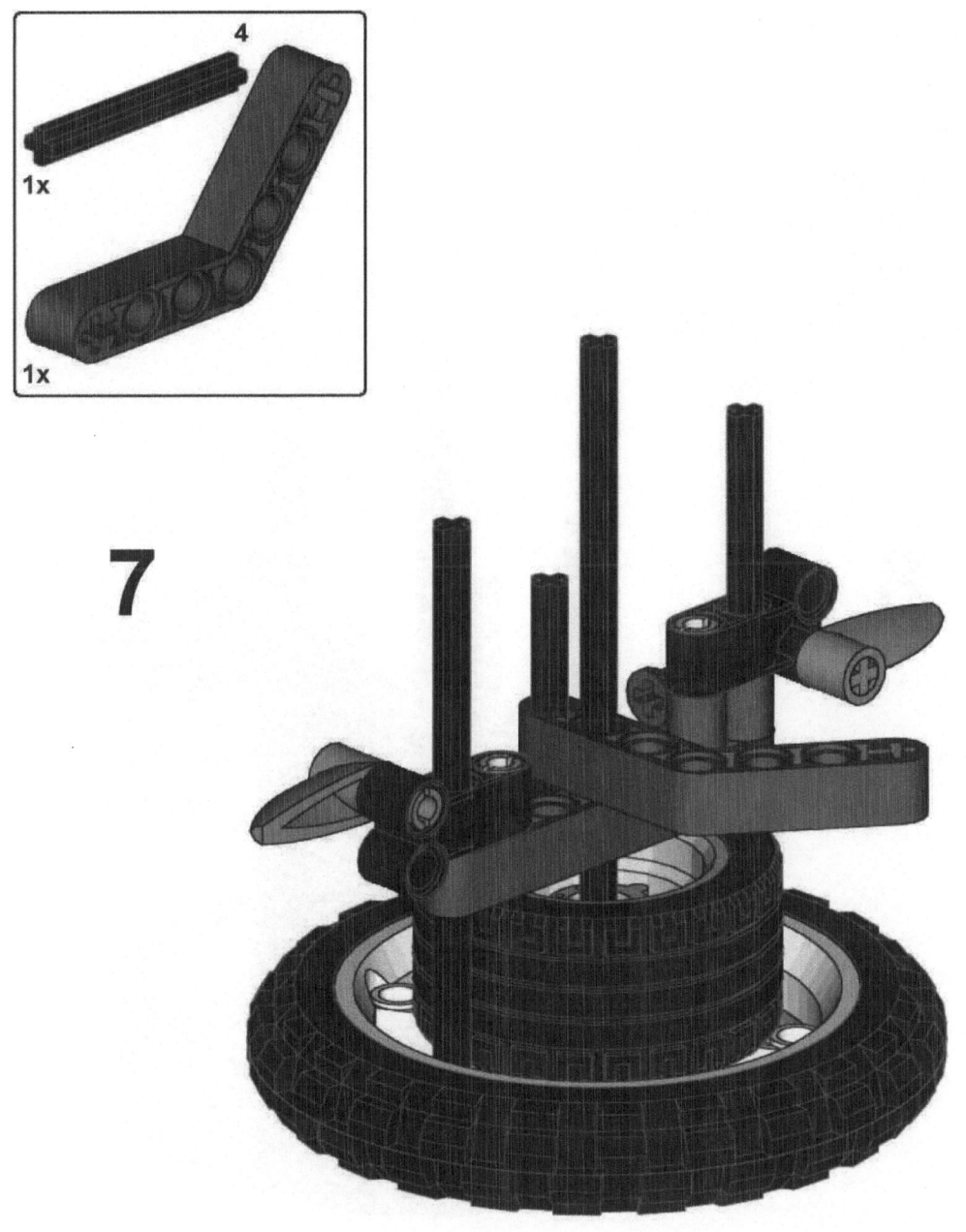

7

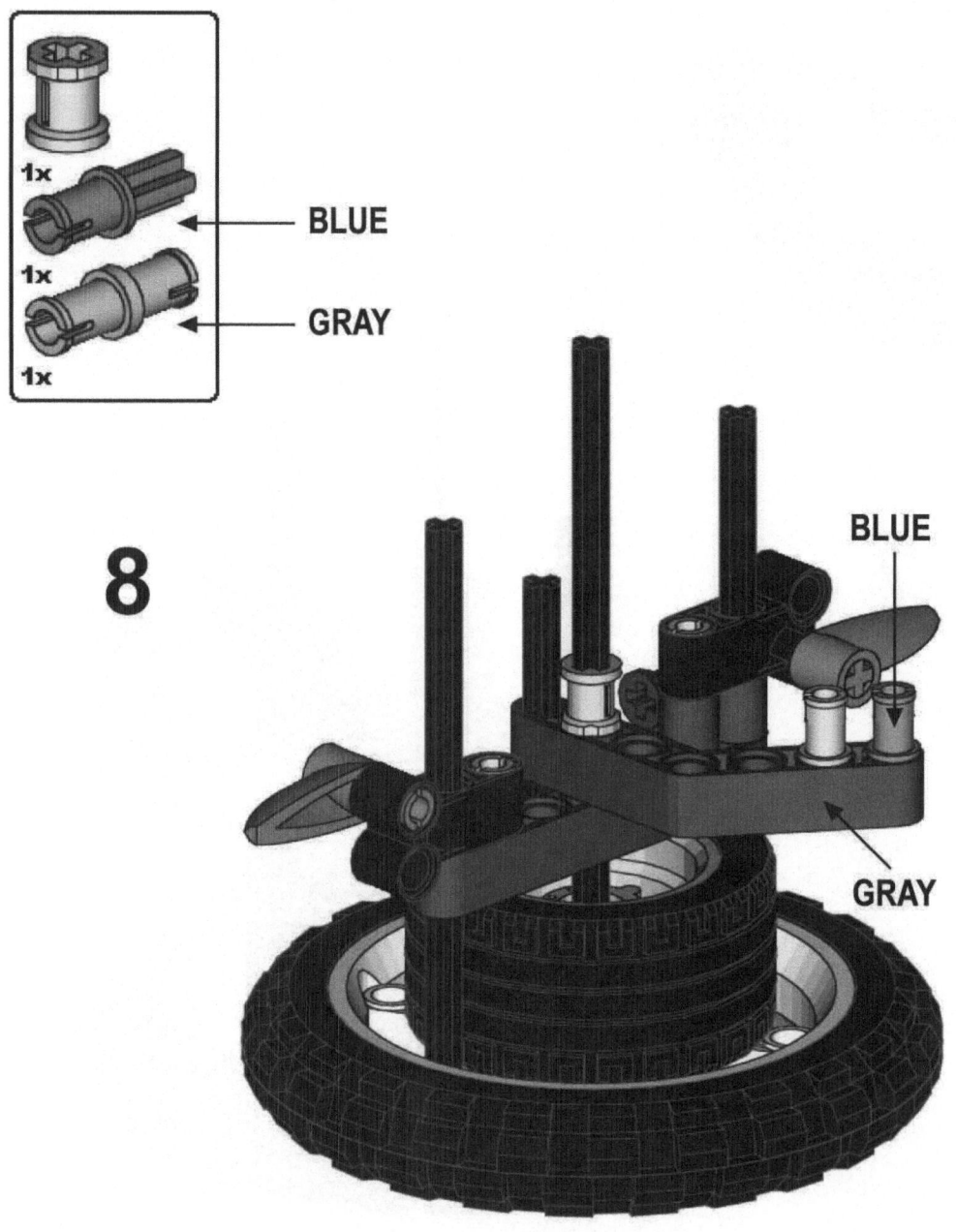

8

BLUE

GRAY

9

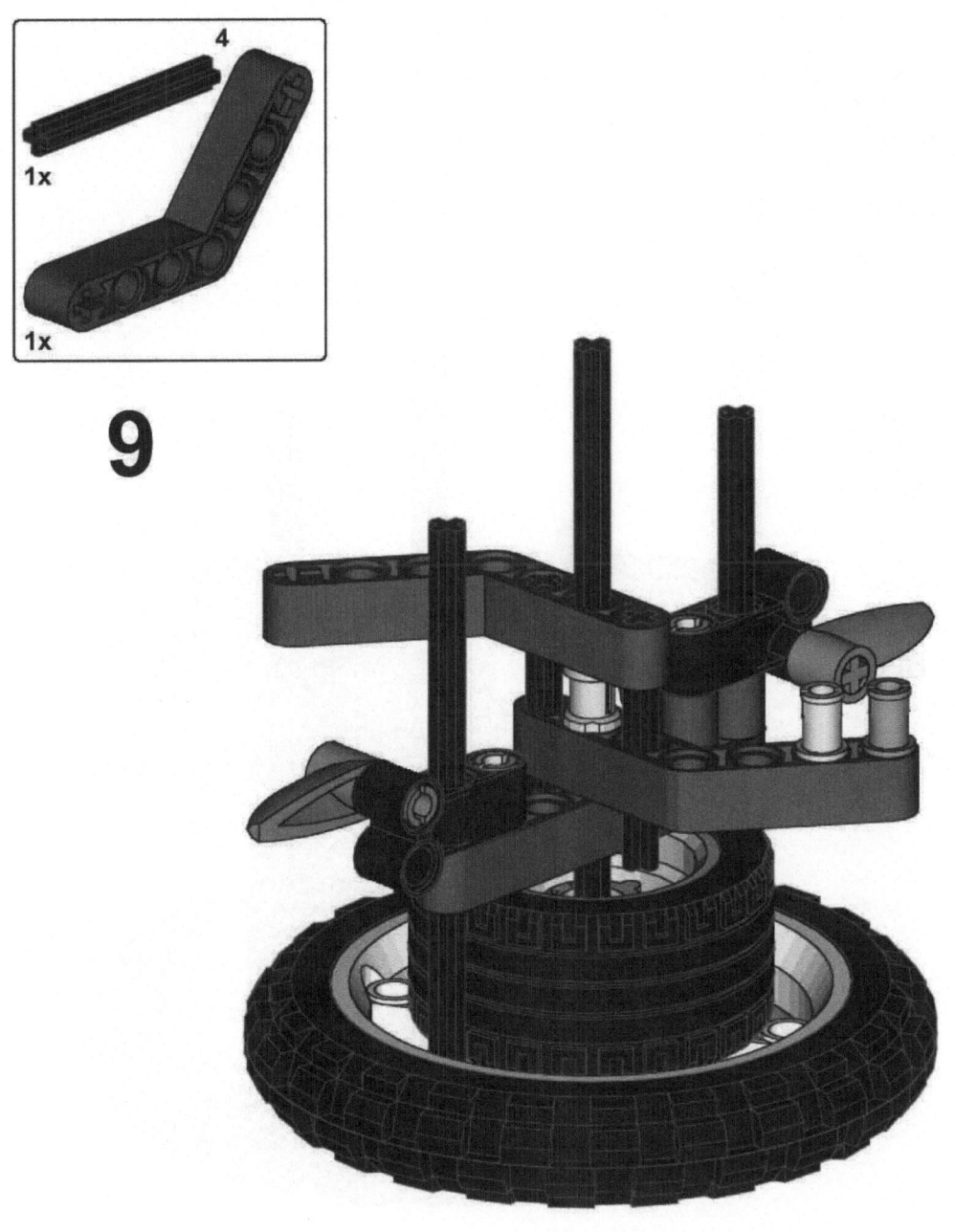

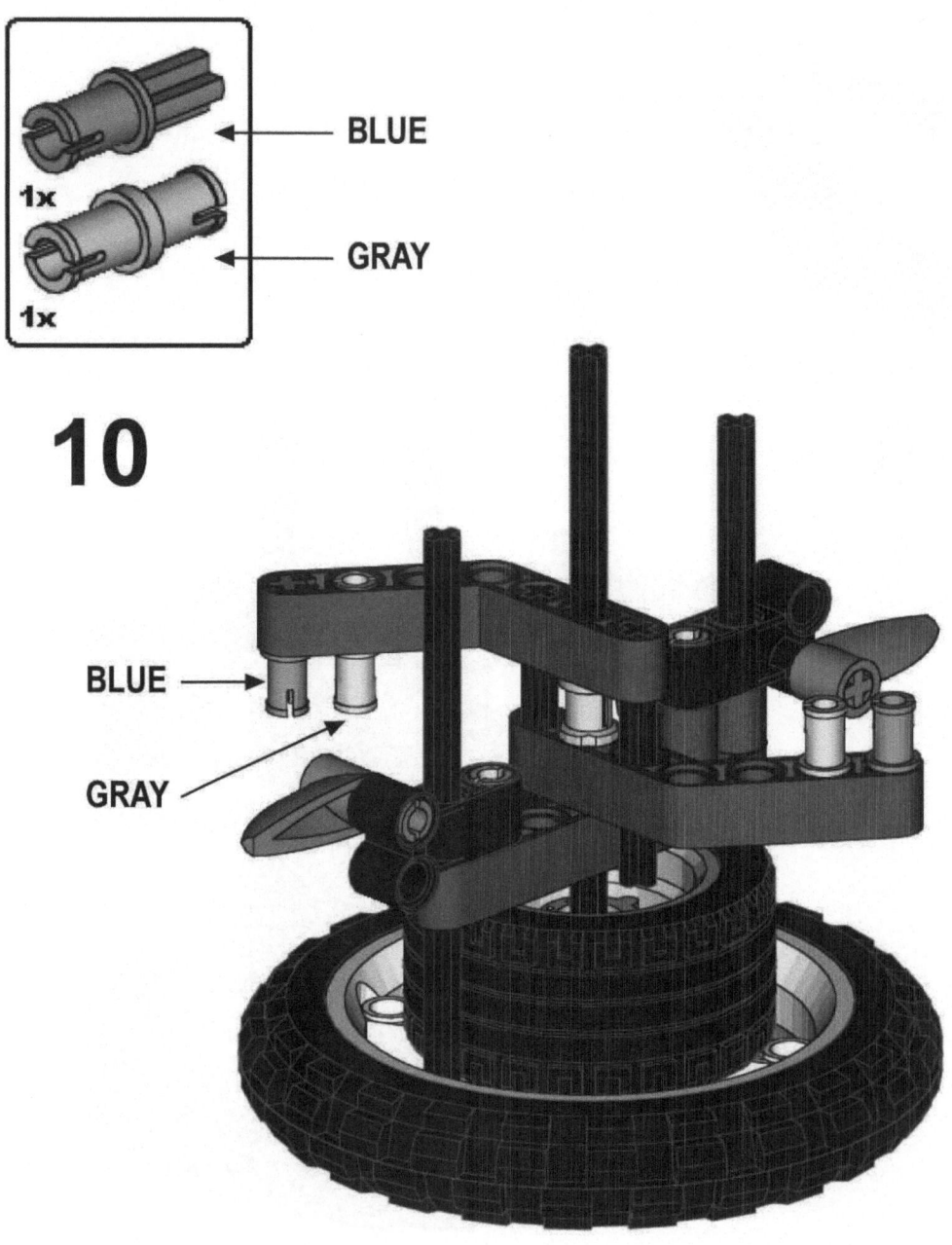

BLUE

1x

GRAY

1x

10

BLUE

GRAY

11

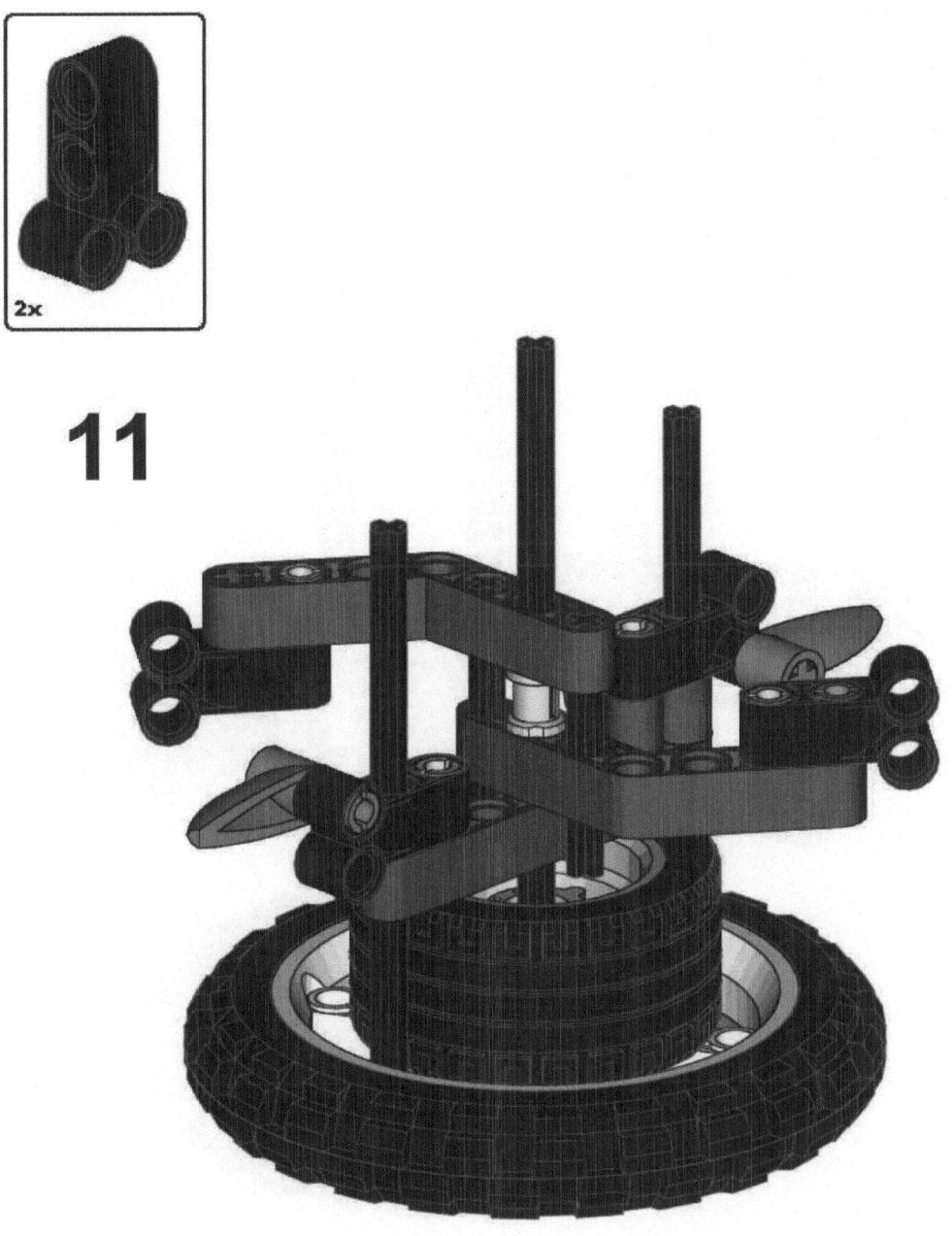

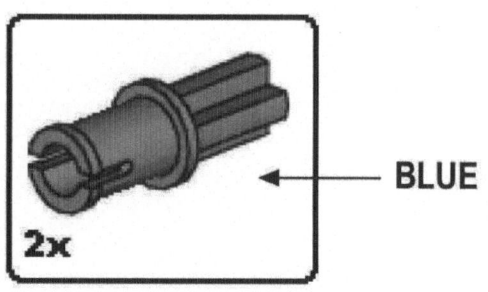

BLUE

2x

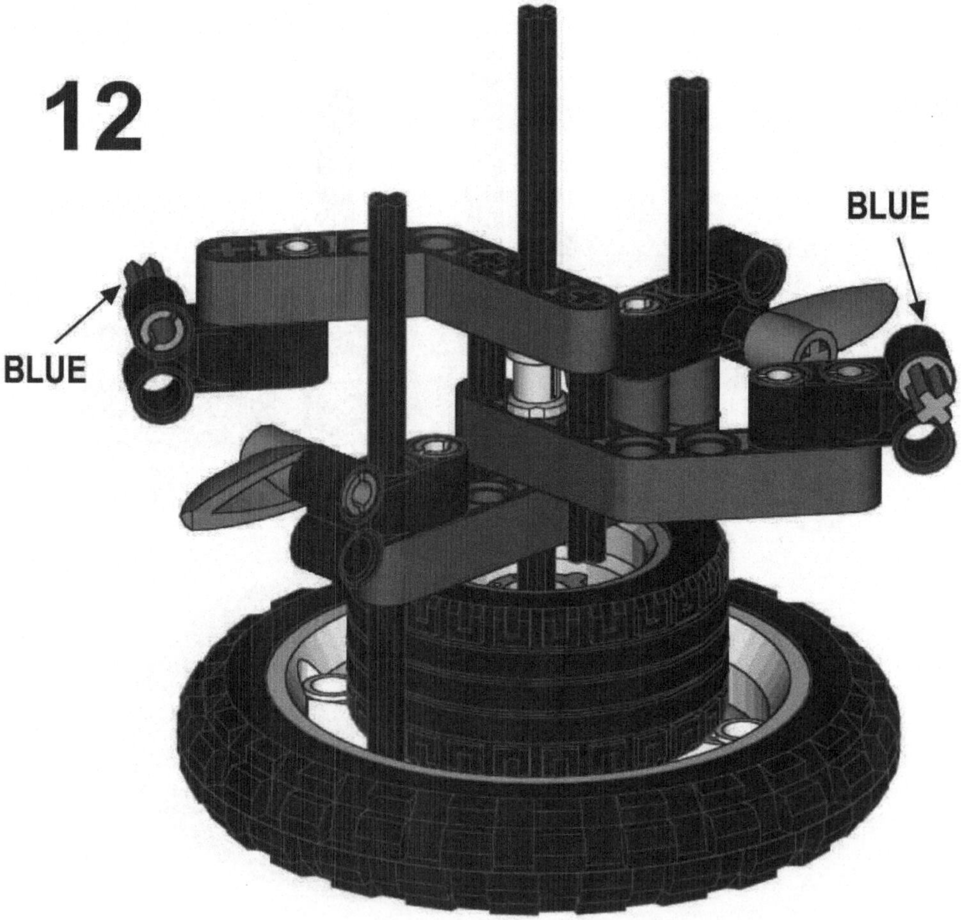

12

BLUE

BLUE

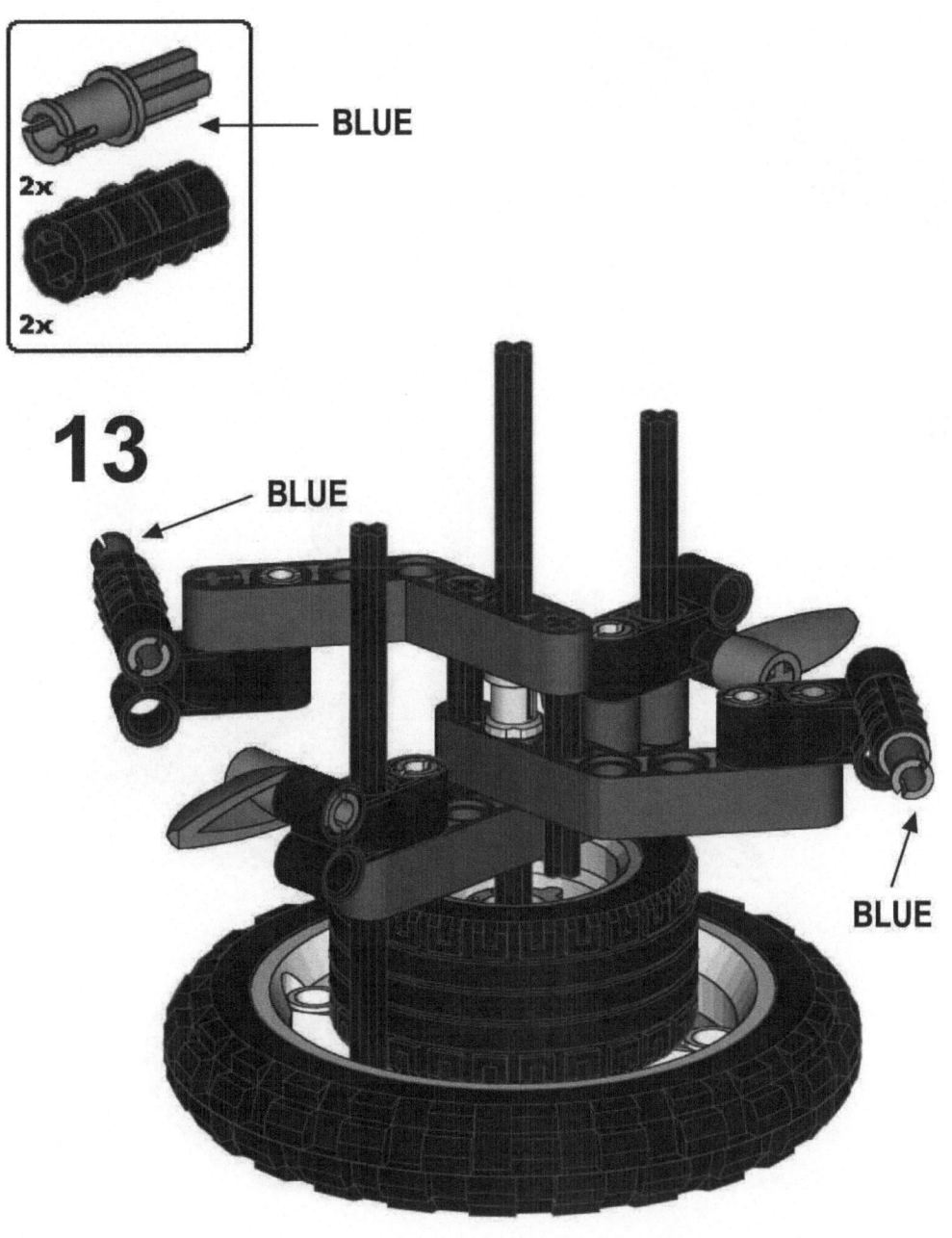

BLUE

2x

2x

13

BLUE

BLUE

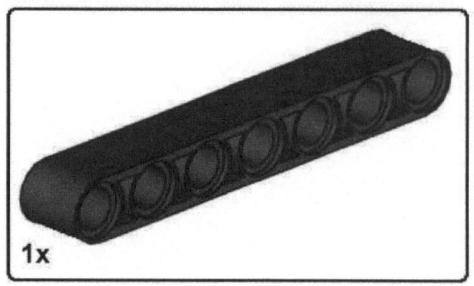

1x

14

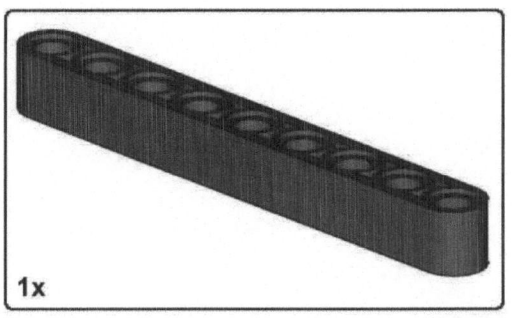

1x

15

GRAY

2x

1x

16

GRAY

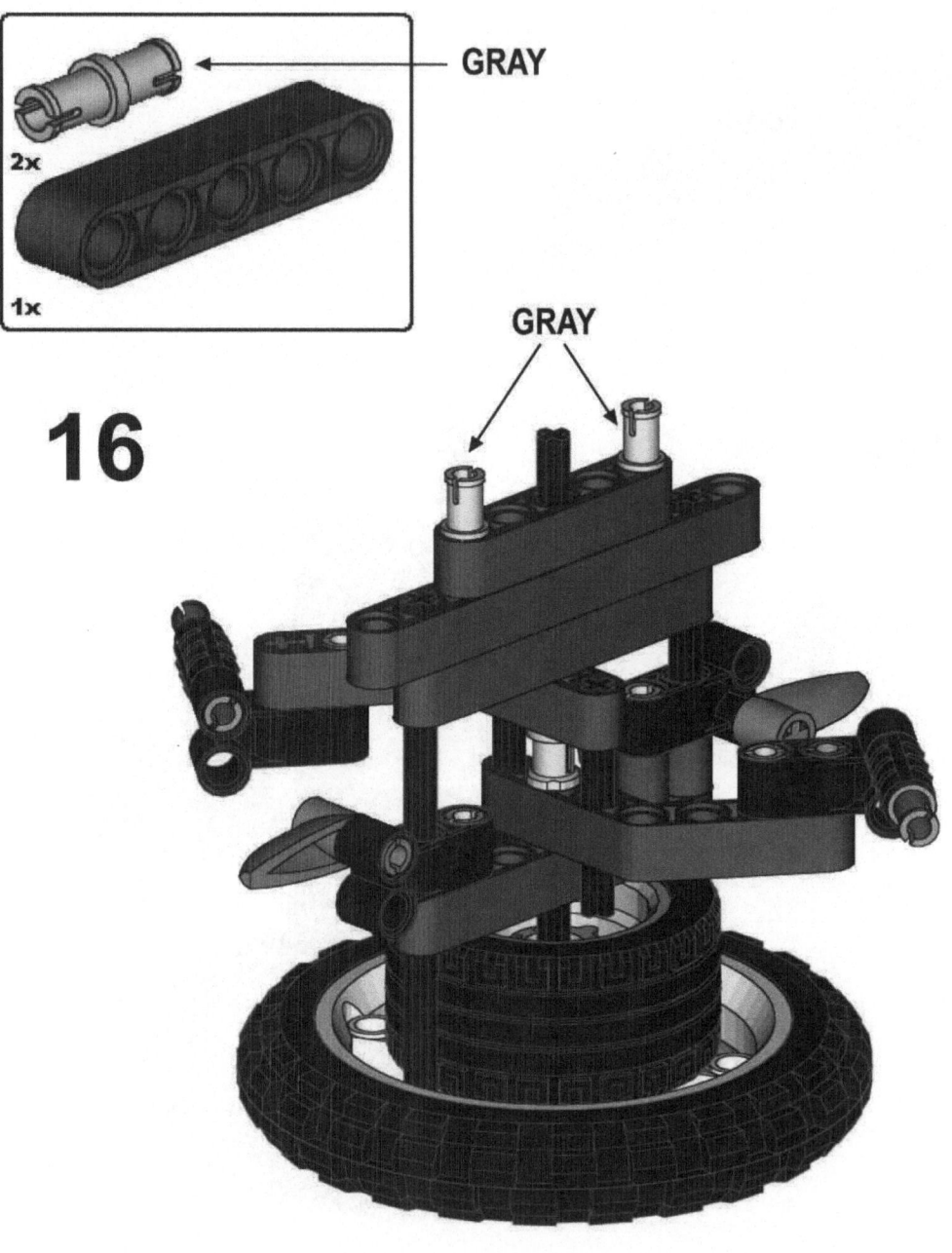

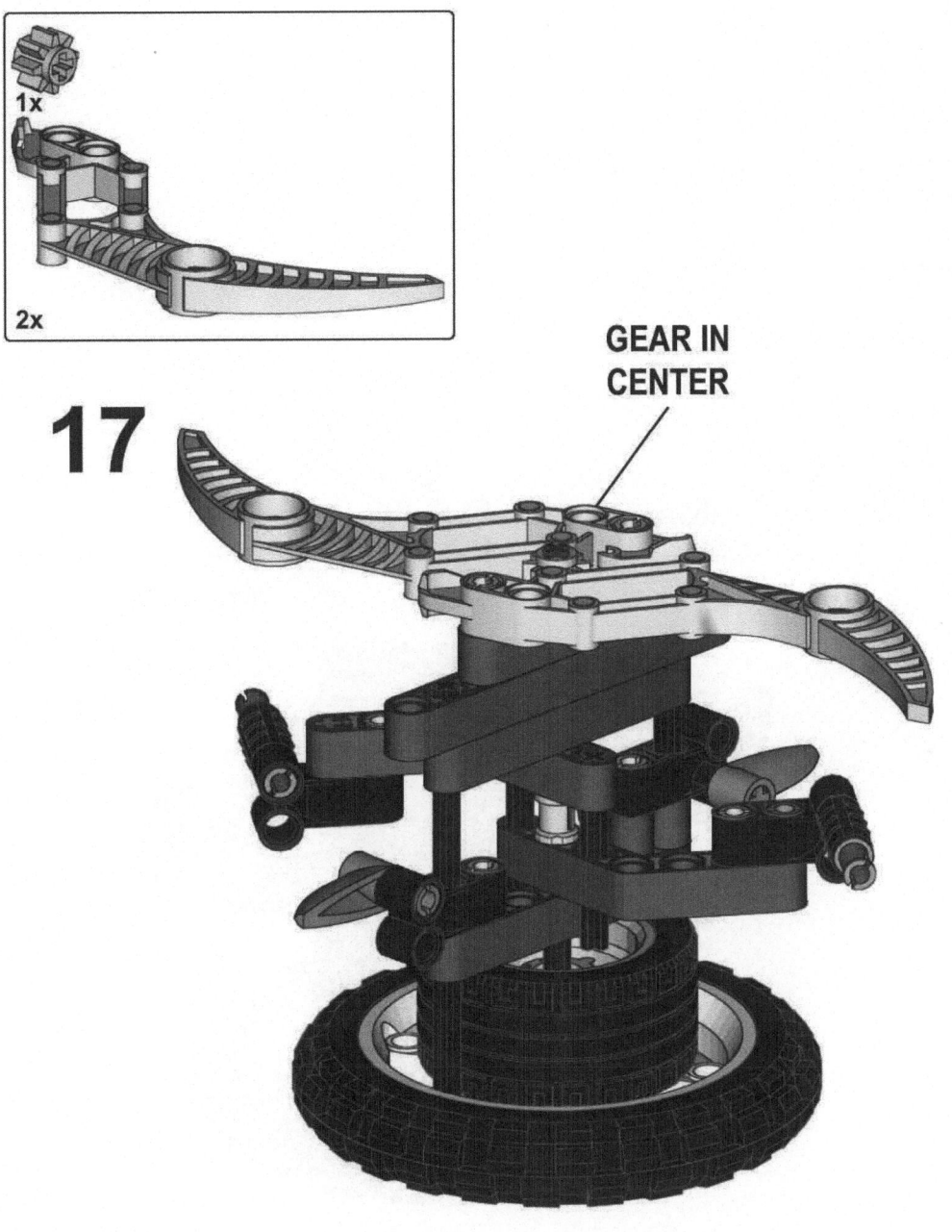

17

GEAR IN
CENTER

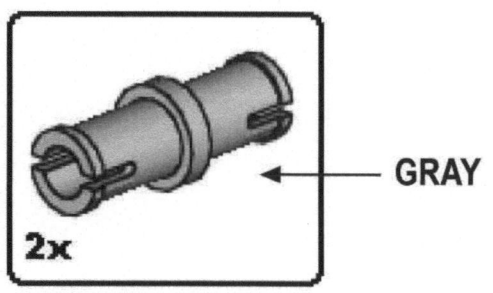

GRAY

2x

18

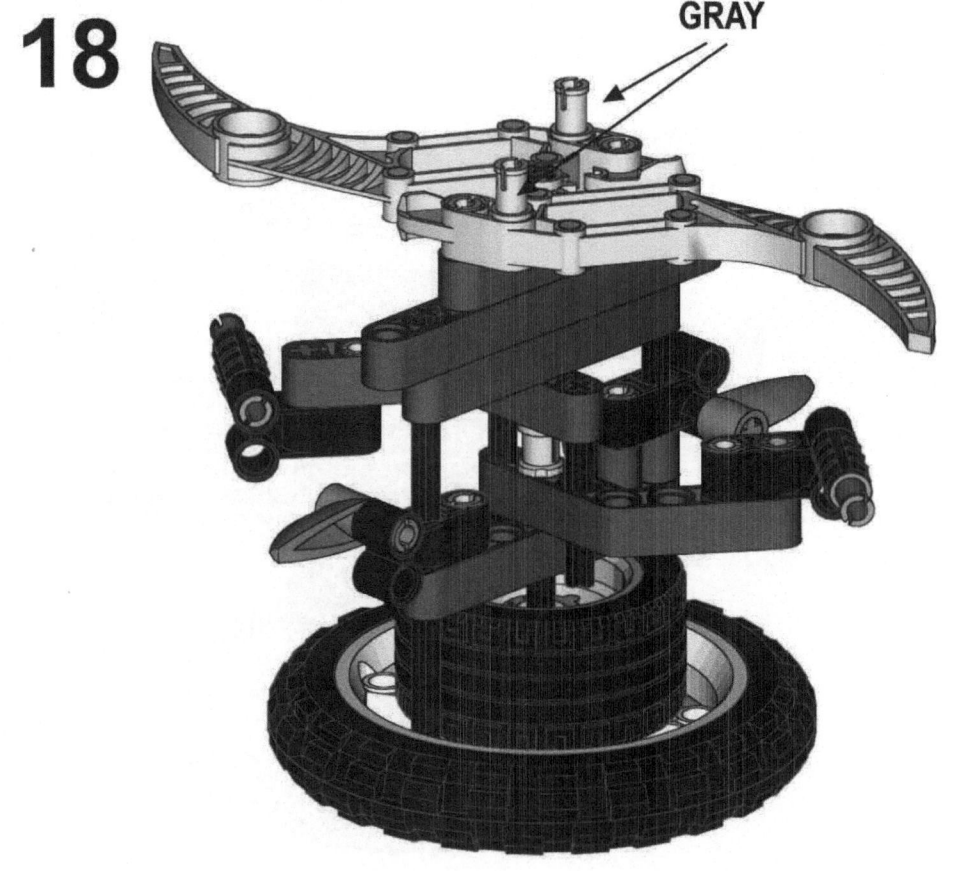

GRAY

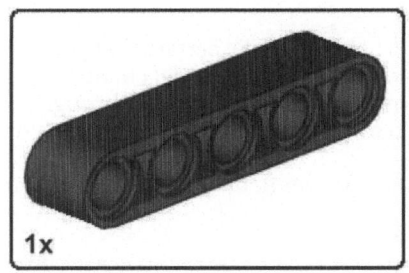

19

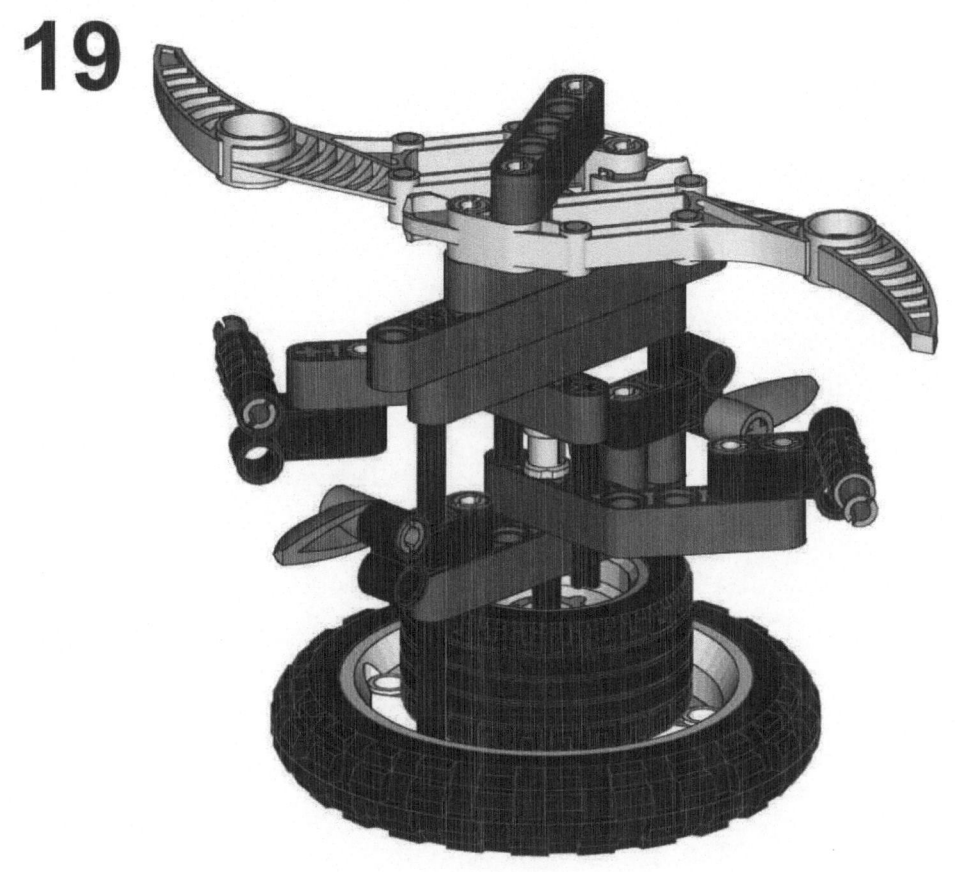

Summary

The models are a critical component to all of the challenge in this book, so make certain that you follow them carefully and build them exactly as described in this chapter and the other model chapters. The models work and have been tested, so if you find a model isn't working properly, go back through the building instructions and verify your assembly. You may find that a single piece has been inserted out of place and has caused the model to not function properly.

Now it's time to continue on to Chapter 3 where you'll build the Power Redirect, the second of three models required for Challenge #1. Have fun.

CHAPTER 3

■ ■ ■

Creating the Power Redirect

In Chapter 2, you were given the steps to build the first of three models required for the Mars Base Alpha challenge. In this chapter, you will learn about the second model, the power redirect (PR), also built using pieces from the LEGO Education Resource Set.

Figure 3-1 shows what the completed power redirect will look like.

Figure 3-1. The power redirect (PR) fully assembled

The PR has only one moving part, a swiveling arm with a barrel of circular pieces on one end and two orange tips on the other end. Chapter 5 will provide details on the placement and operation of the PR in the challenge.

Note   You will also find information related to the challenge area; this can be quickly put together with nothing but the three assembled models for the Mars Base Alpha challenge and some tape to define the boundaries. You will also find details on a mat that can be downloaded (for free) and printed out in color or black and white and used as the challenge area.

Tackling the PR Challenge

The PR is not a difficult model to figure out. In the challenge, the pointers (two orange tips) will be pointing toward the right, and your robot must rotate the PR so that it points to the left. This will simulate redirecting the power obtained from the solar array (see Chapter 4) to Mars Base.

There is no danger to your robot from this device other than the possibility that it could get caught up in the structure. Notice also that there are two large rubber tires in front of the PR; should your robot bump into these, it is possible that the robot could be bumped off course or briefly caught, because the tire surface can be tacky and stick to plastic surfaces.

You will definitely want to take a measurement or two that will allow you to determine the proper height (from the ground) that your robot must touch the circular barrel or the pointer ends on the PR's arm. You must also be able to properly program your robot so that it moves the arm from right to left. It doesn't matter if your robot approaches the PR from the front or back as it will need to perform the same action to turn the arm from left to right. But if you examine the PR carefully, you'll find that the orange pointer ends and the circular barrel on the other end of the arm aren't always at the same height. This is something to consider when programming your robot.

Taking on One Challenge at a Time

After you've assembled the power redirect, consider building a small, single-task robot as was suggested in Chapter 2. You can use this test robot to try to find a solution to the PR challenge. Rather than worrying about building and programming a robot to interact with the power interrupt device and the power redirect, simply try to build one to solve the PR problem.

In this instance, you've got to find a way to get your robot from the PID location to the PR location.

Note   The Mars Base Alpha challenge may not make sense to you yet, so feel free to jump ahead to Chapter 5 to read about the rules of the challenge. You'll find that your robot must follow steps in order as it interacts with the three models. In the full challenge, your robot will first attempt to disengage the PID before approaching the power redirect. For testing purposes on this second challenge item only, place your robot where the PID would be located and assume that it has successfully disengaged the PID and is now ready to find the PR.

Use a combination of measurements on the mat (or floor) from PID to PR as well as visual cues, such as colors or lines on the mat or tape (you are allowed to place tape on the floor; there's no rule against it). A sensor or two can give your robot the ability to navigate the challenge area and find the PR before attempting to move the arm.

Building the Power Redirect

In this section, you will find the building instructions for the power redirect. Each image shows the pieces you will need to locate in the LEGO Education Resource Set and their quantities. You will also see where these pieces are to be placed.

If you are uncertain about the placement of a piece in a figure, jump ahead to the next image or even go back to a previous image to make certain you've built the model correctly so far. Examine the figures carefully, and you should be able to correctly identify the pieces and their final location on the model.

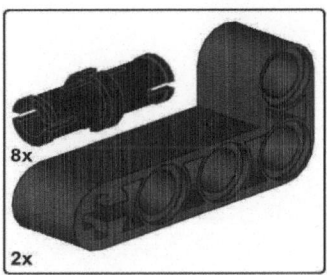

1

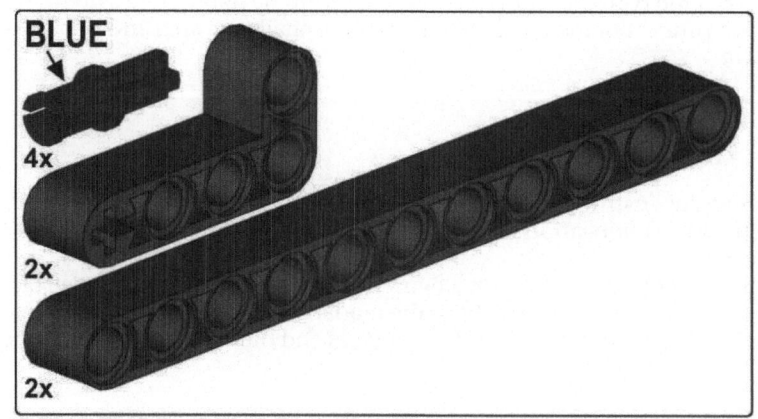

2

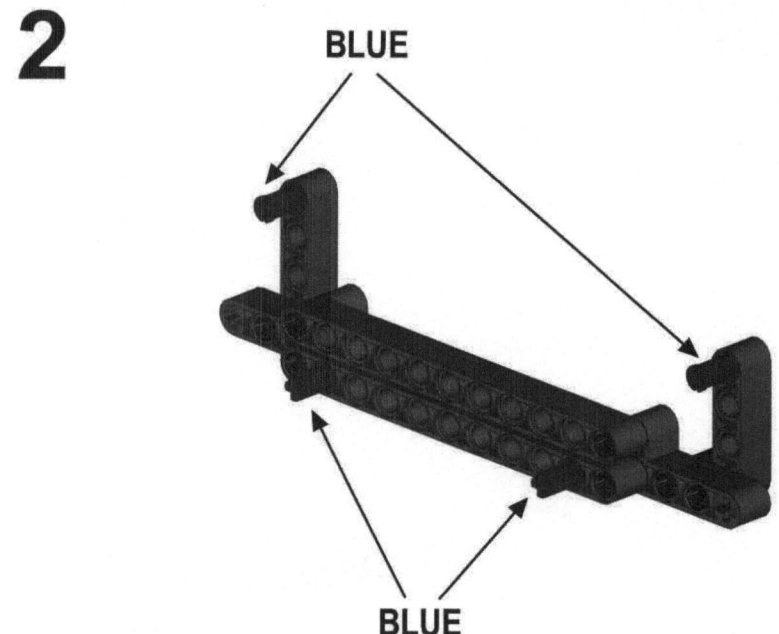

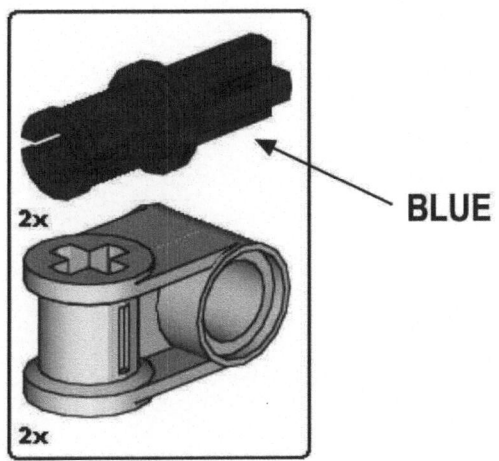

BLUE

2x

2x

3

BLUE

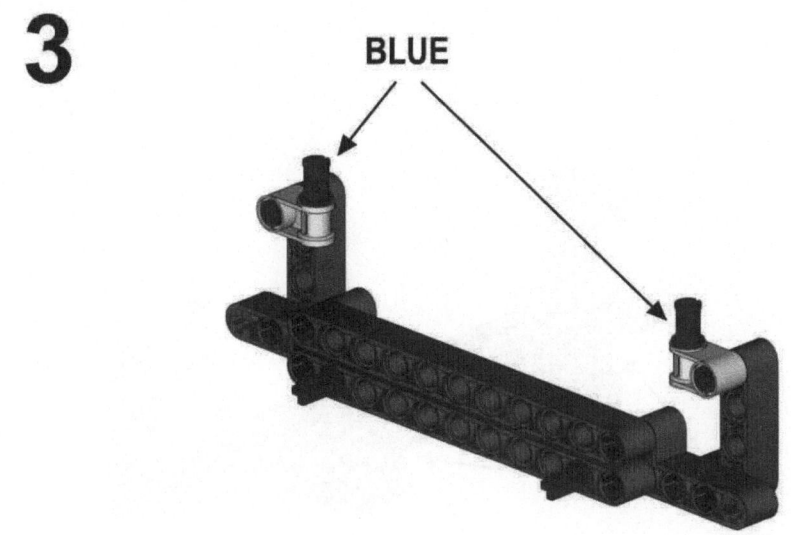

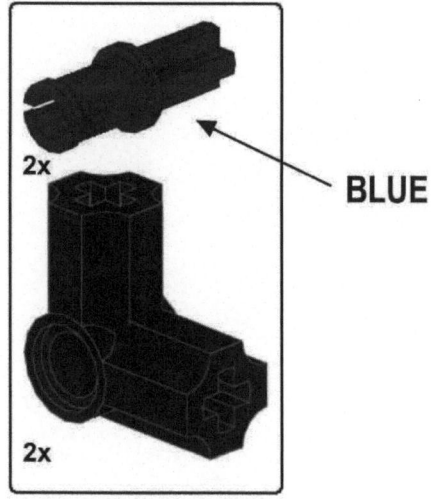

BLUE

2x

2x

4

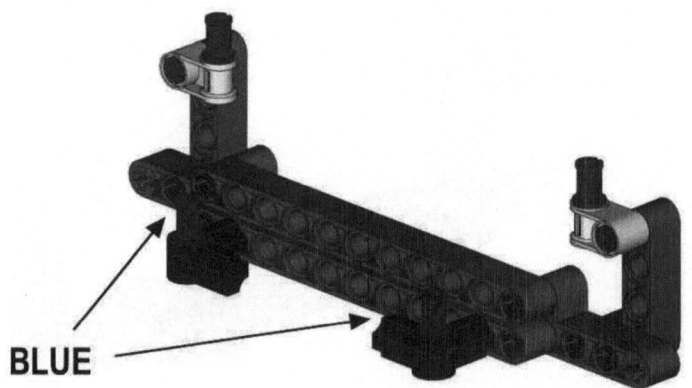

BLUE

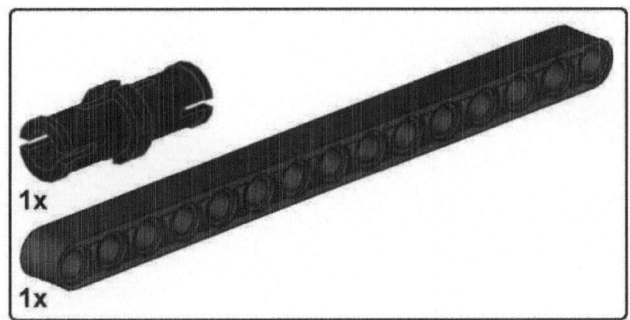

5

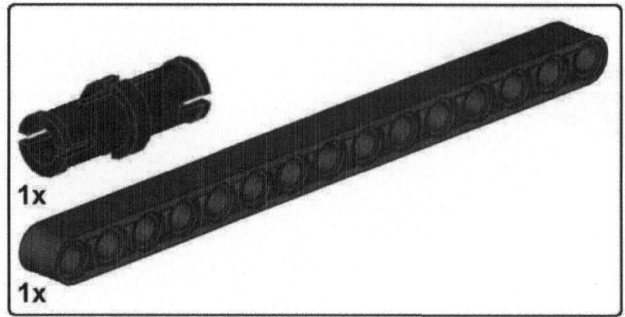

6

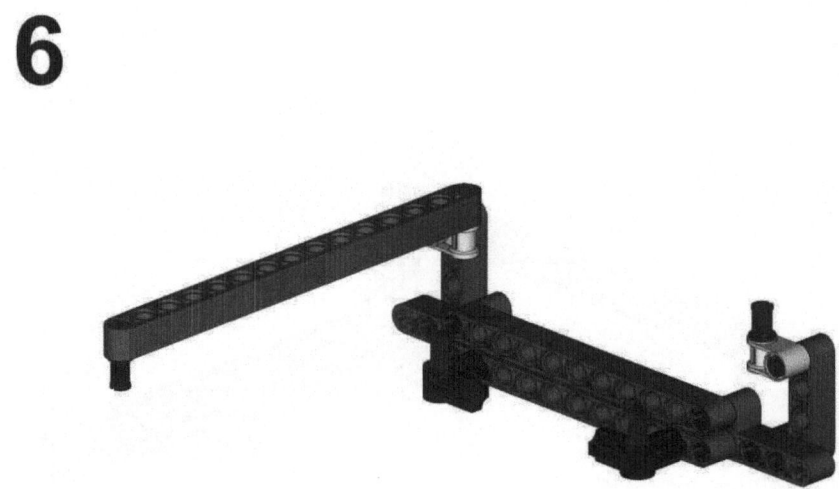

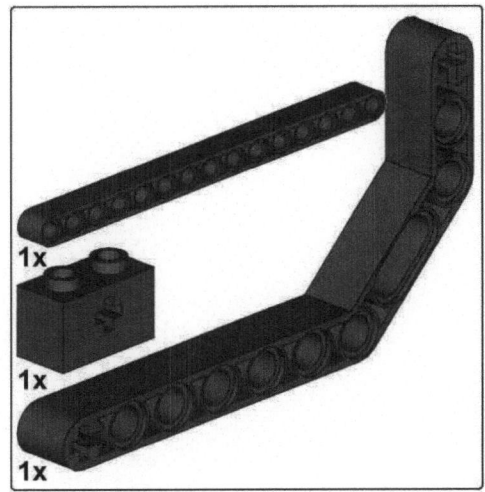

7

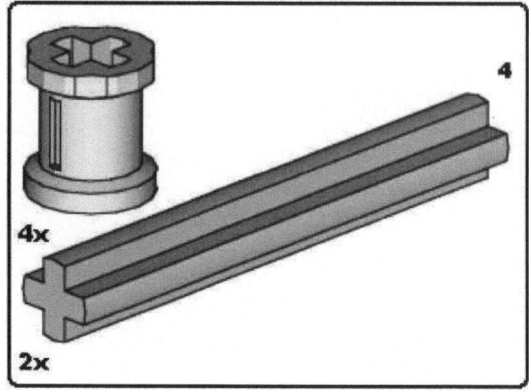

8

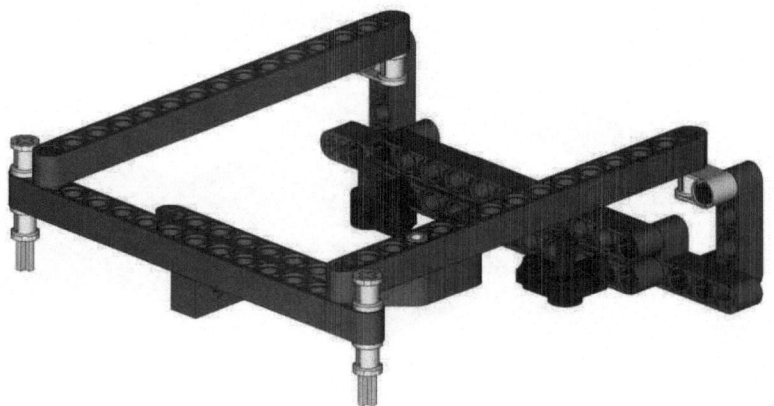

9

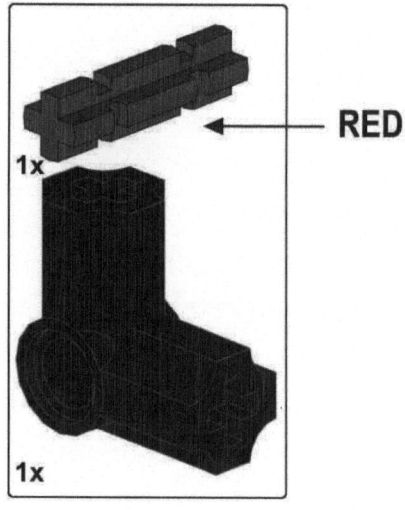

RED

1x

1x

10

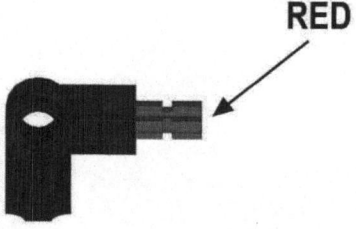

RED

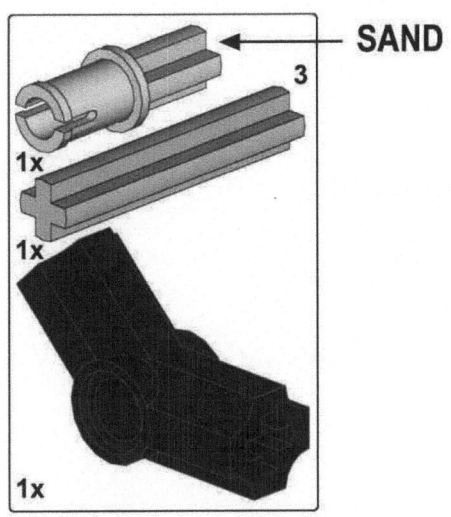

11

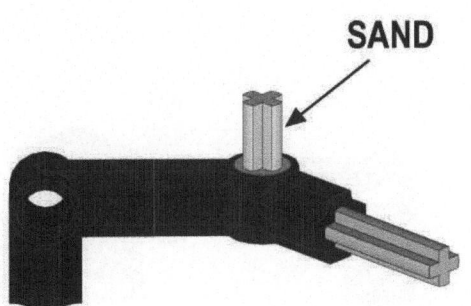

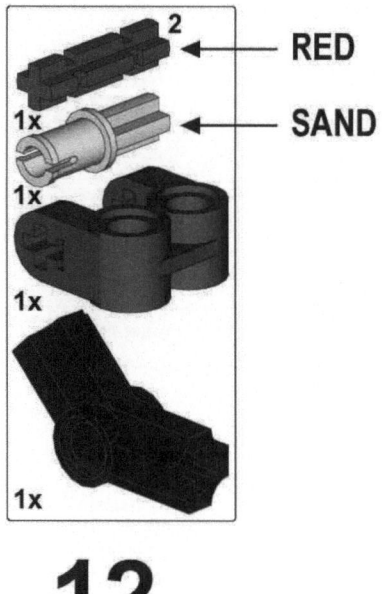

RED

SAND

12

SAND

RED

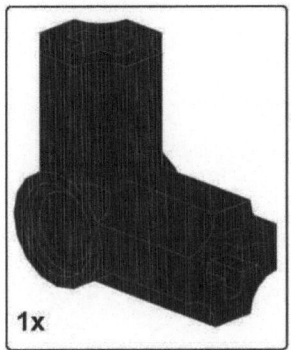

13

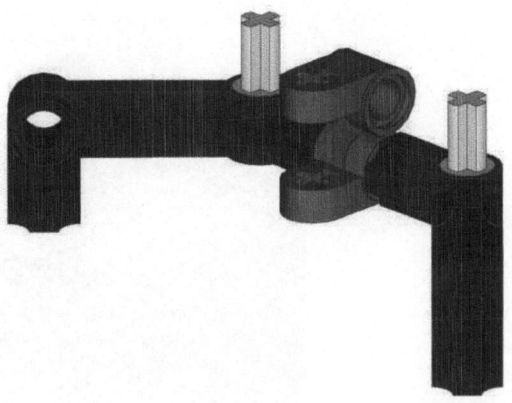

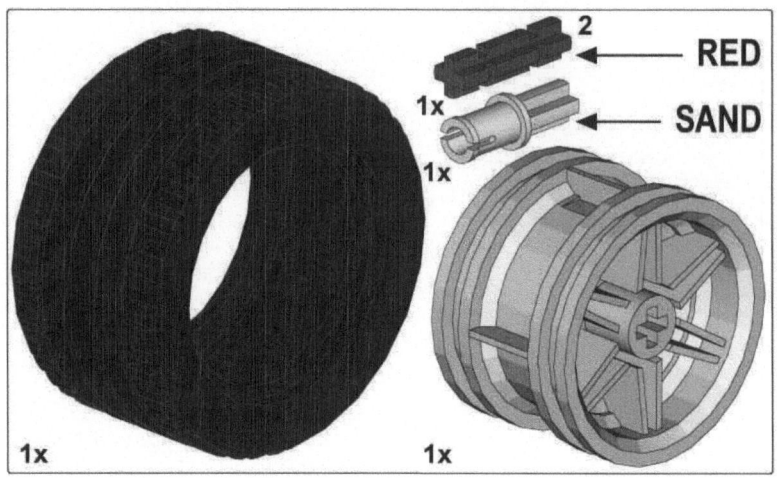

14

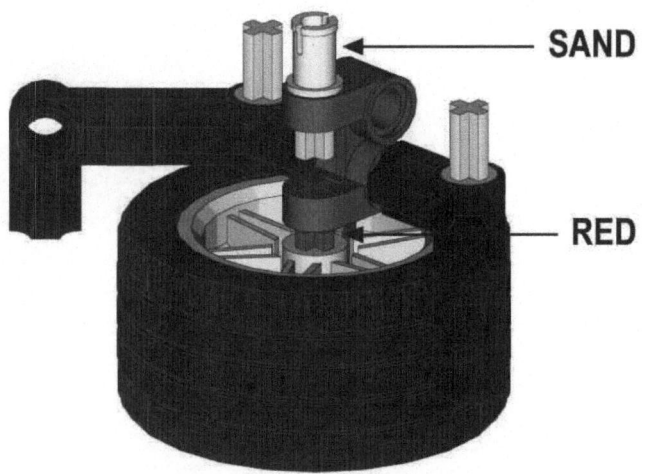

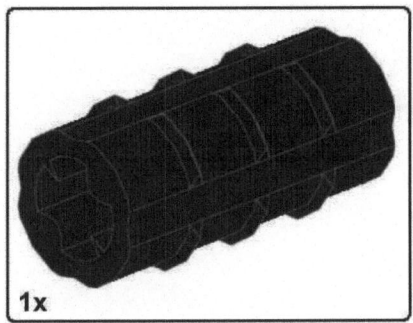

1x

15

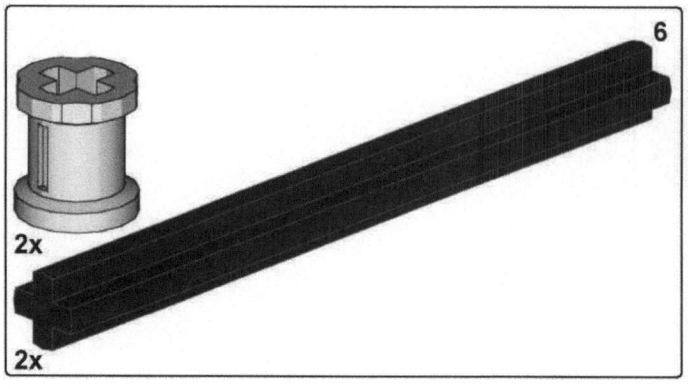

16

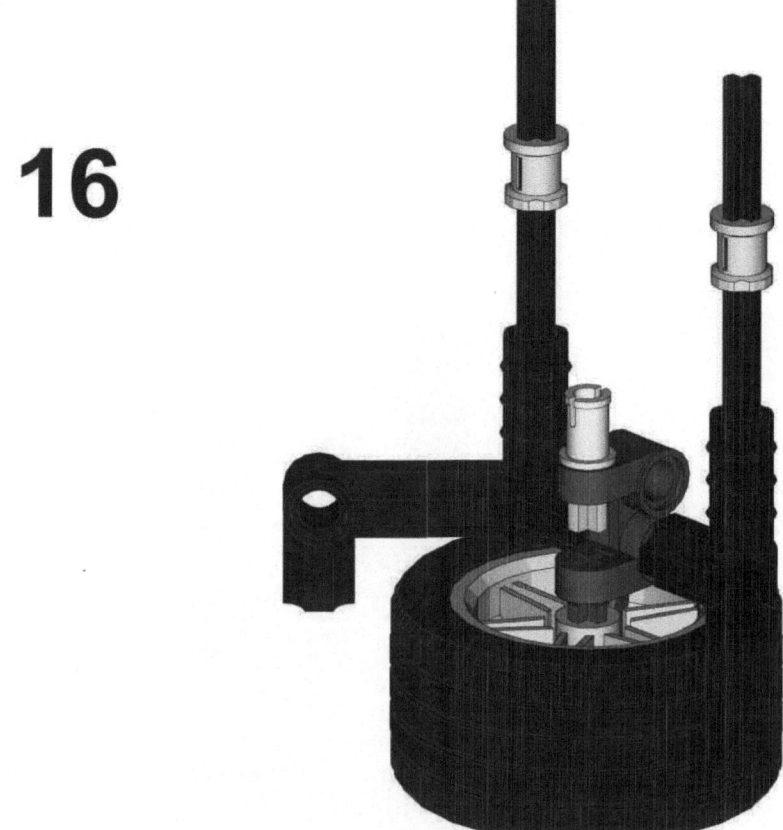

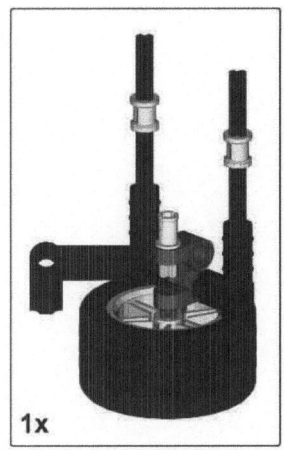

17

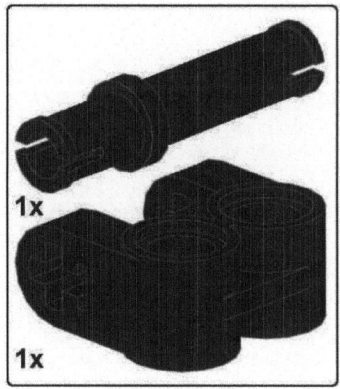

18

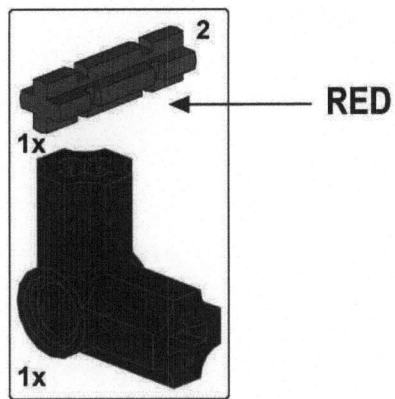

RED

19

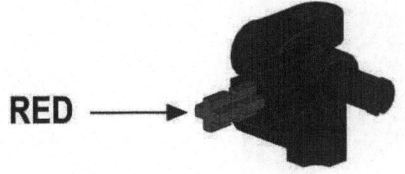

RED

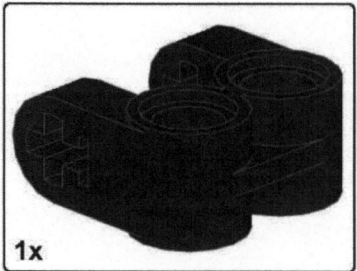

20

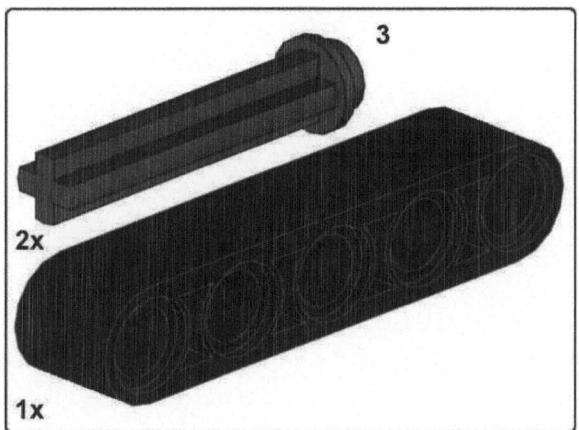

21

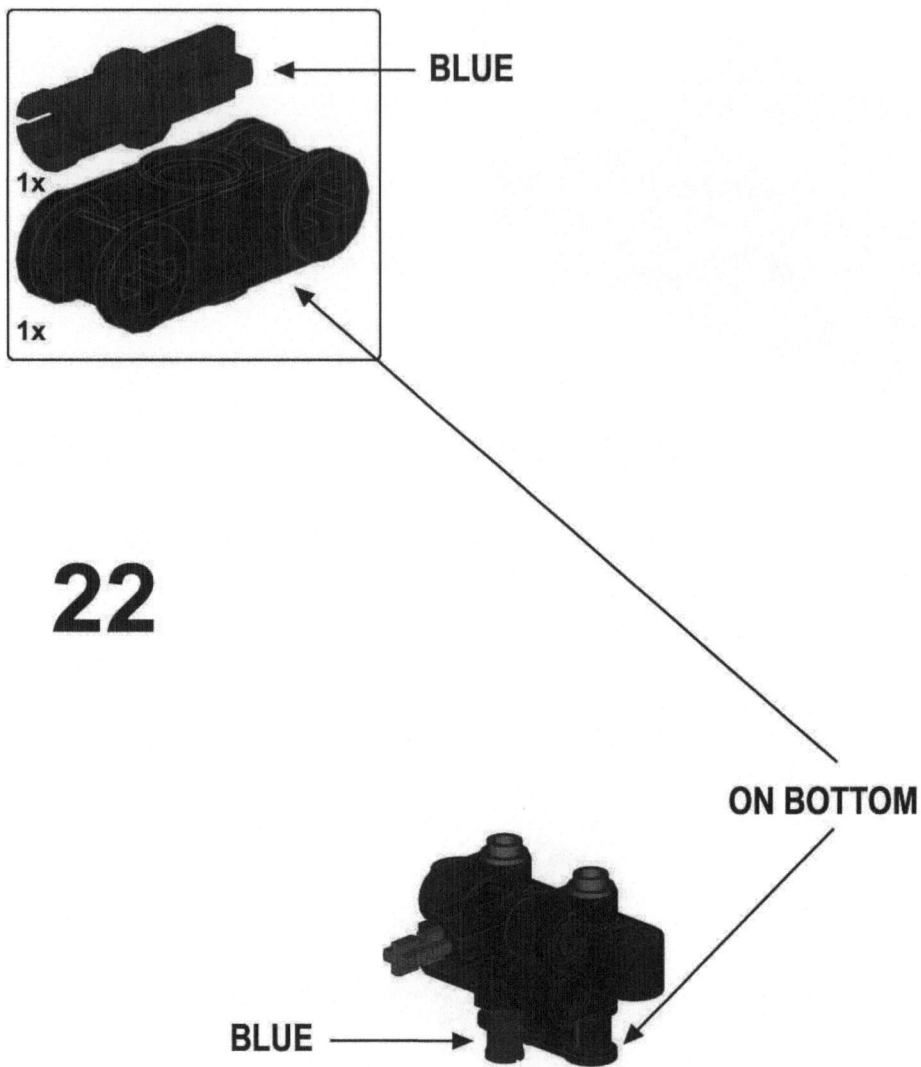

22

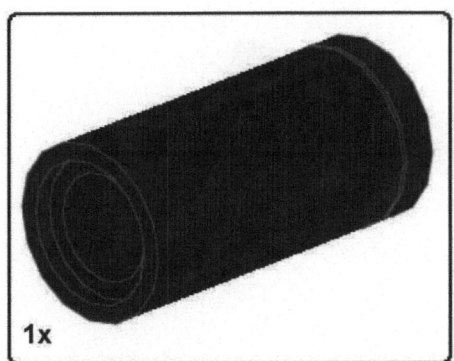

23

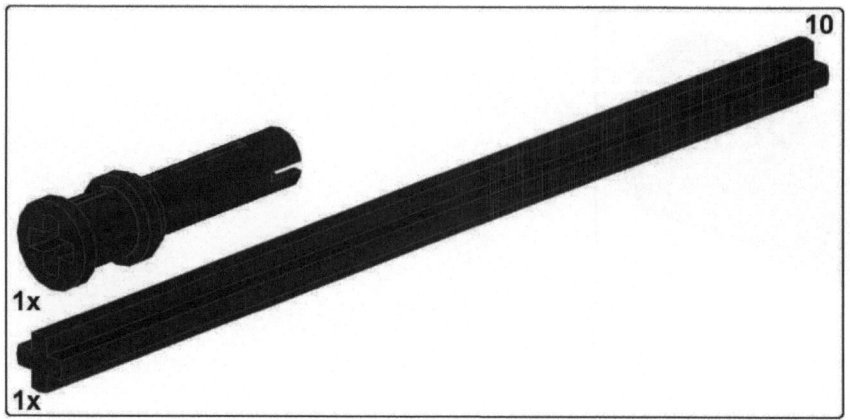

24

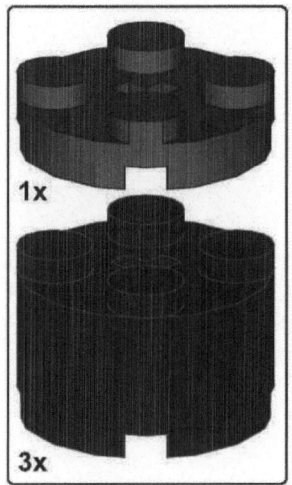

25

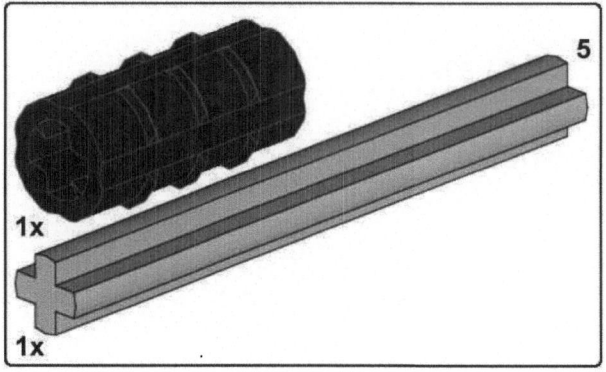

26

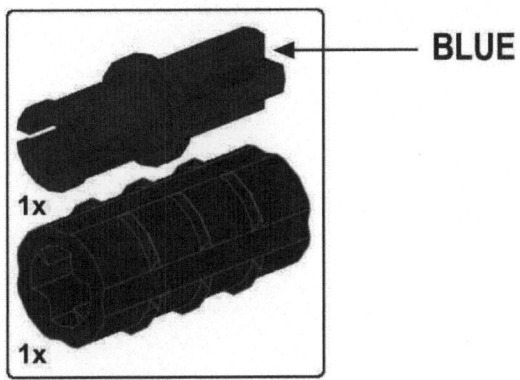

BLUE

1x

1x

27

BLUE

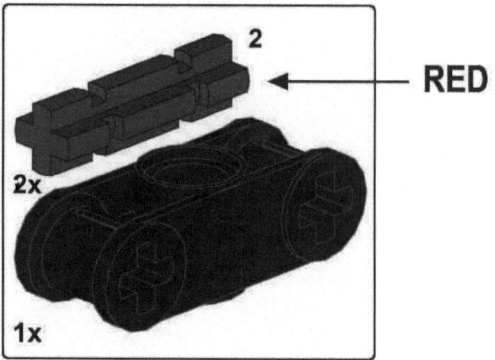

28

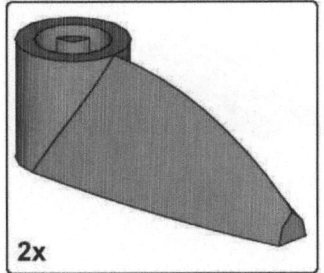

2x

29

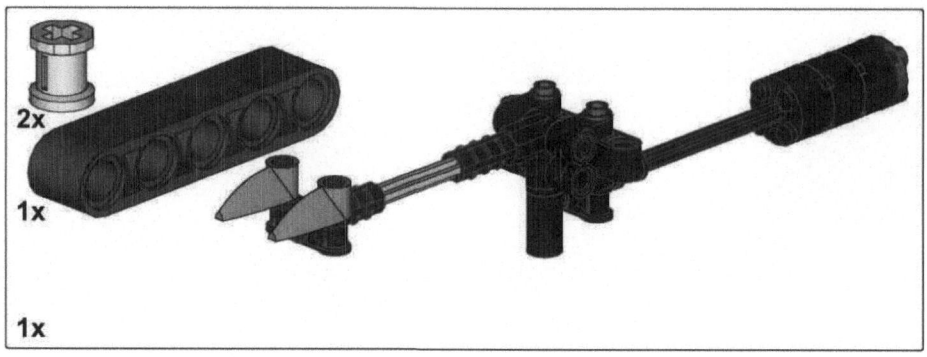

30

Summary

Hopefully you're beginning to see how the models will work in the Mars Base Alpha simulation. You've now completed two of the three models required to run the first challenge, but as you work through the book, keep in mind that you can always modify the models if you like. Make them easier or harder for your robot to use. Or remove them altogether if they're simply too difficult.

Up next in Chapter 4, you'll build the final model, the Solar Collector. After that model is built, move on to Chapter 5 and learn how to set up the first challenge and run it.

CHAPTER 4

■ ■ ■

Creating the Solar Collector

In Chapters 2 and 3, you were given the steps to build the first two (of three) models required for the Mars Base Alpha challenge. In this chapter, you will learn about the final model, the solar collector (SC) built using pieces from the LEGO Education Resource Set.

Figure 4-1 shows what the completed power redirect (PR) will look like.

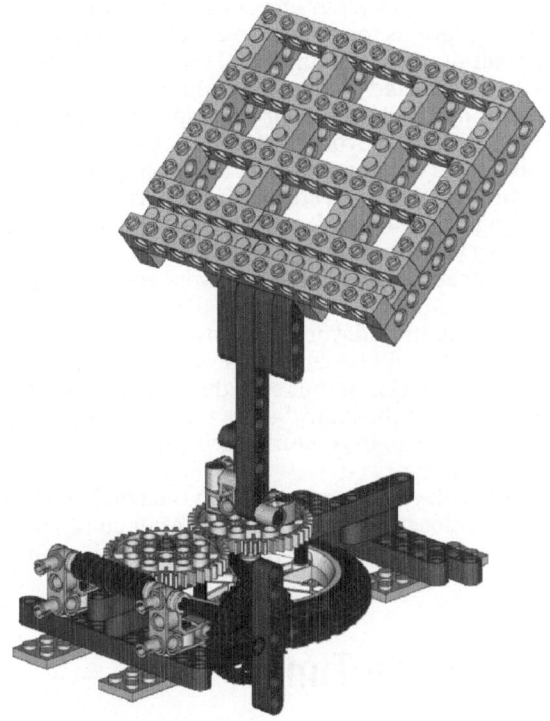

Figure 4-1. The solar collector fully assembled

The solar collector has two moving parts: one is an X-shaped crank that must be turned by your robot, and the other is the solar cells mounted near the top that will rotate as your robot turns the crank. Chapter 5 will provide details on the placement and operation of the SC in the challenge.

Note   You will also find information related to the challenge area; this can be quickly put together with nothing, but the three assembled models for the Mars Base Alpha challenge and some tape to define the boundaries. You will also find details on a mat that can be downloaded (for free) and printed out in color or black and white and used as the challenge area.

Tackling the SC Challenge

The solar collector is, by far, the most difficult challenge your robot faces in the Mars Base Alpha challenge. You must design and program your robot in such a way that it can locate the X-shaped crank and then turn it a sufficient number of rotations to turn the solar cells on top 180 degrees (to face the reverse of its starting position). This will simulate the turning of the solar collectors to face the sun so that power can begin to be collected and directed to Mars Base Alpha.

There is no danger to your robot from this device; your robot may touch any part of this model but do be careful not to allow your robot to either tip over the SC or get caught in any of the moving parts.

First, you'll want to get your robot from the power redirect (see Chapter 3) to a position near the solar collector where the robot can turn the X-shaped crank. You can use a combination of programming using measurements and programming using sensor feedback to get the robot to the proper position; consider using visual cues on the mat (if you are using it) such as colors and lines. If you are not using the mat, consider using colored tape placed on the floor to help your robot navigate to the proper position.

Given that the X-shaped crank must be turned, you must now consider how you will turn that crank. The most likely method (but certainly not the only one) is to use a motor, placed somewhere on the robot's body, that can be programmed to rotate and thus spin the crank. This will require some testing of various mechanisms that can either lock on to the crank before turning it or perform incremental rotations (similar to using a slot screwdriver where you turn once, pull out the screwdriver before inserting again and turning again…and so on).

Whatever your solution, when finished, your robot must turn the crank enough times to rotate the solar cell panel on top 180 degrees (or as close to that as possible). After completing the rotation of the SC, your robot must then find the power interrupt device (PID), approach it again, and reengage it (see Chapter 2).

Taking on One Challenge at a Time

The solar collector model is the most difficult model your robot will encounter in the Mars Alpha Base challenge. Not only must your robot find the SC and correctly align itself so that it can turn the crank but it must continue to rotate the crank until the SC is turned 180 degrees.

As was suggested in Chapters 2 and 3, the best way to tackle this challenge is to build a single robot that will find and turn the SC's crank. Don't worry about the PID or the PR devices; build and program a robot to successfully complete the solar collector challenge, and everything else will fall into place.

When you're finished, remember also that your robot must find and approach the power interrupt device one more time; your robot will need to reengage the PID before the challenge is completed.

Note   The Mars Base Alpha challenge may not make sense to you yet, so feel free to jump ahead to Chapter 5 to read about the rules of the challenge. You'll find that your robot must follow steps in order as it interacts with the three models. In the full challenge, your robot will first disengage the PID before approaching the PR. After moving the arm of the PR, it will find and turn the SC. Only after turning the SC will the robot move back to the PID and reengage it. For testing purposes, place your robot where the PR would be located and assume that it has successfully move the PR's arm (left to right) and is now ready to find the SC.

You'll once again need to use measurements on the mat (or floor) from PR to SC as well as visual cues such as colors or lines on the mat (or tape, which you are allowed to place on the floor). Sensors can be used to give your robot the ability to navigate the Challenge Area and find the SC before attempting to rotate the crank.

Building the Solar Collector

In this section, you will find the building instructions for the solar collector. Each image shows the pieces you will need to locate in the LEGO Education Resource Set and their quantities. You will also see where these pieces are to be placed.

If you are uncertain about the placement of a piece in a figure, jump ahead to the next image or even go back to a previous image to make certain you've built the model correctly so far. Examine the figures carefully, and you should be able to correctly identify the pieces and their final location on the model.

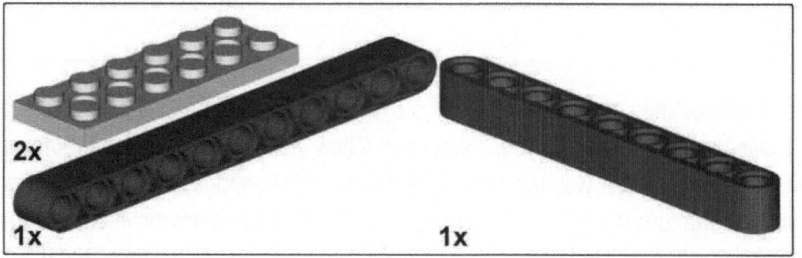

1

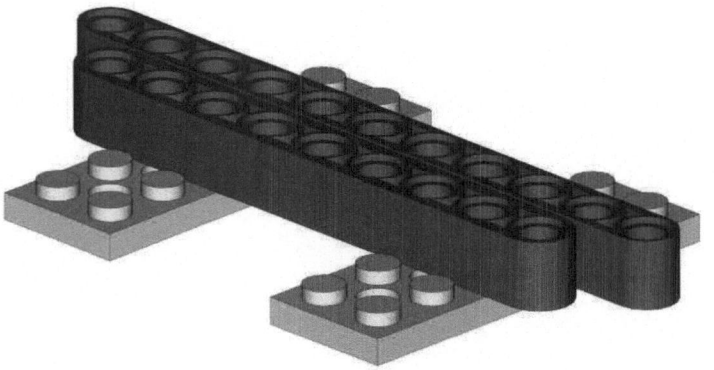

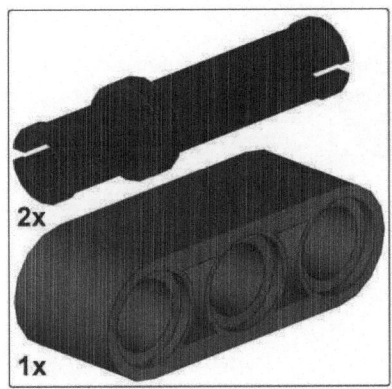

2

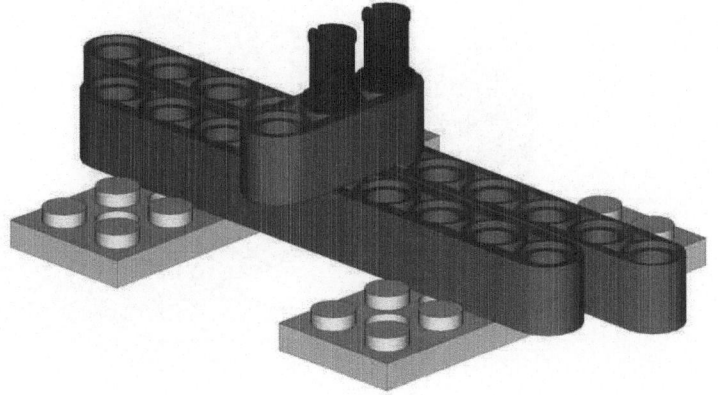

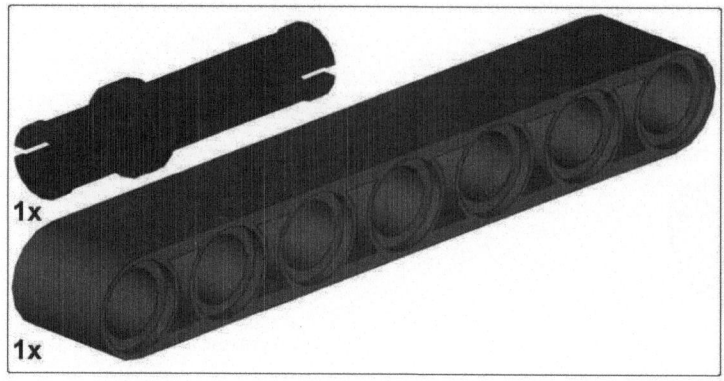

3

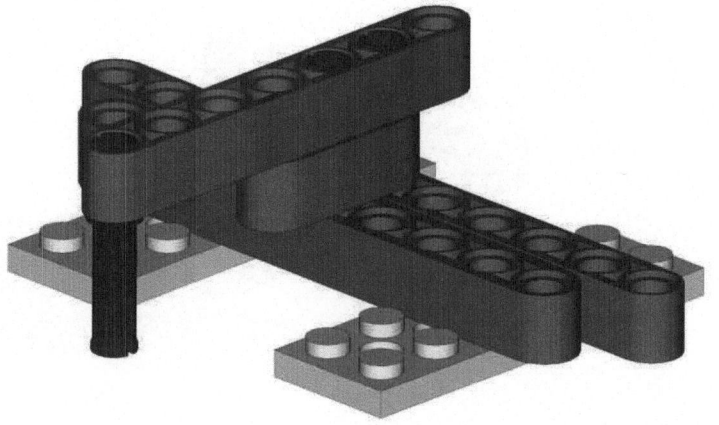

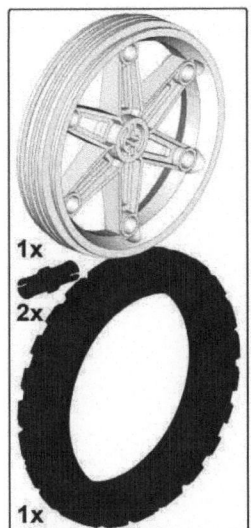

4

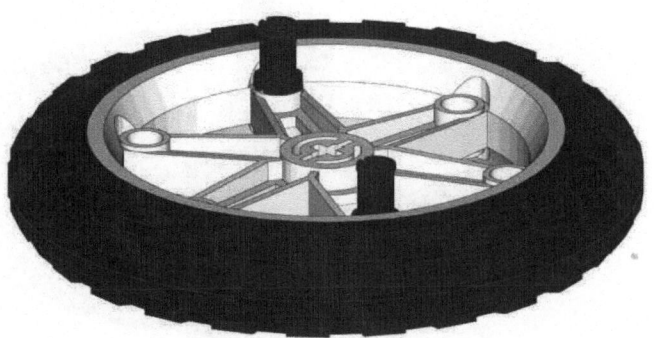

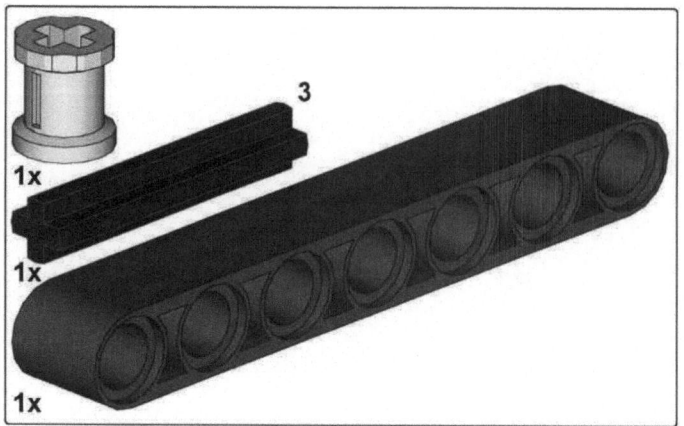

5

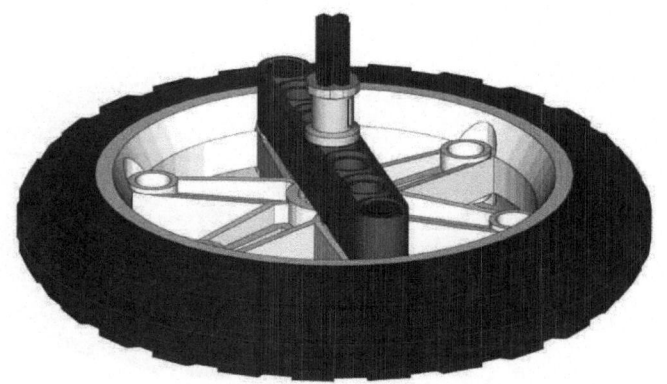

1x

6

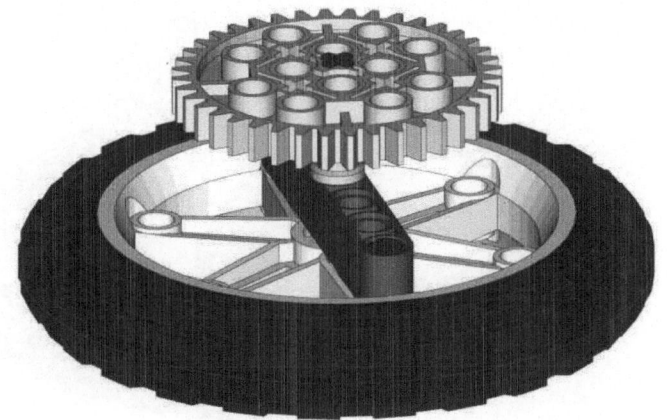

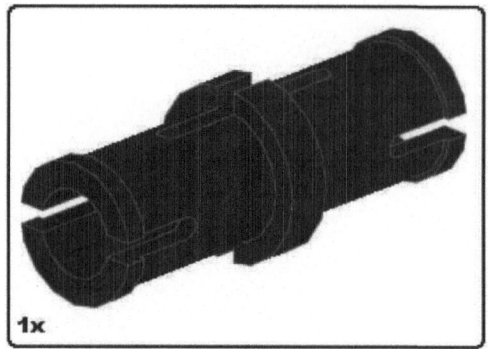

1x

7

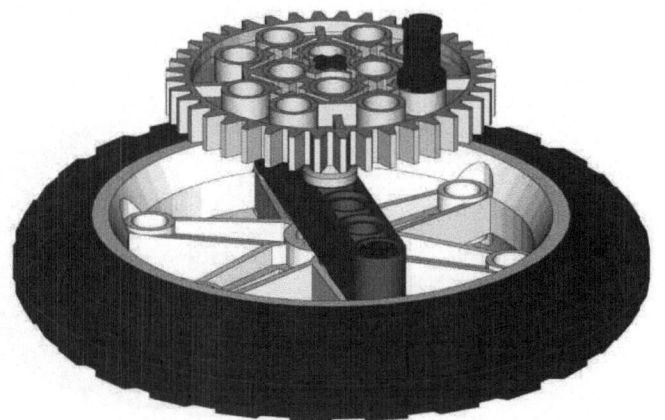

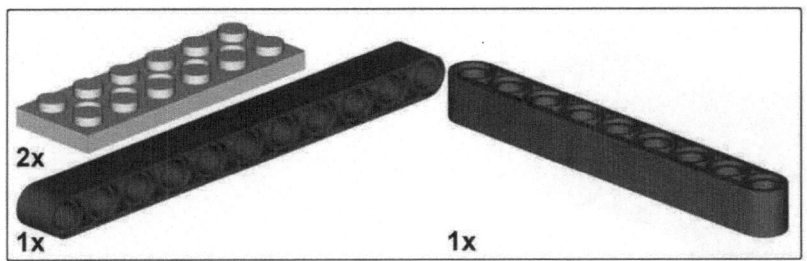

8

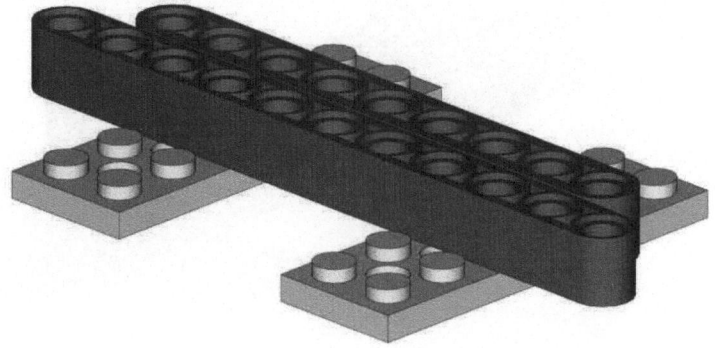

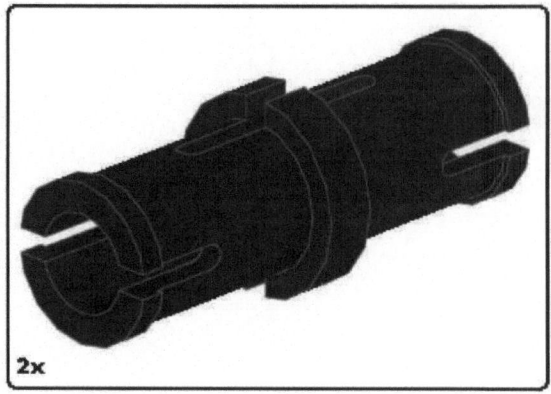

2x

9

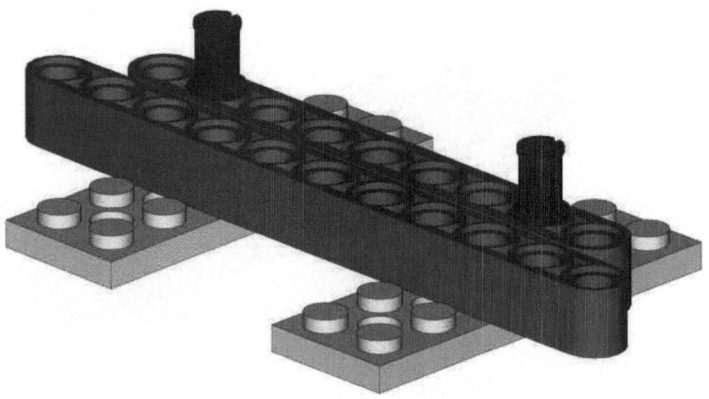

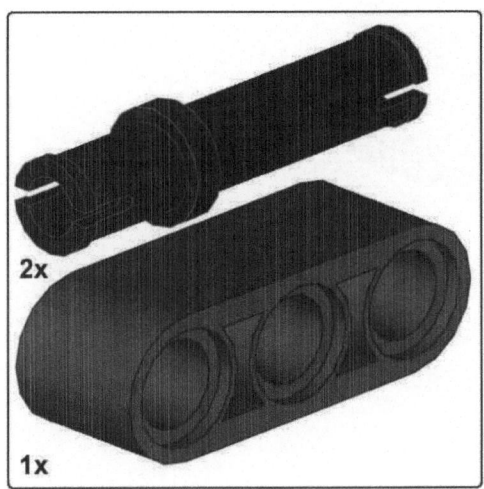

10

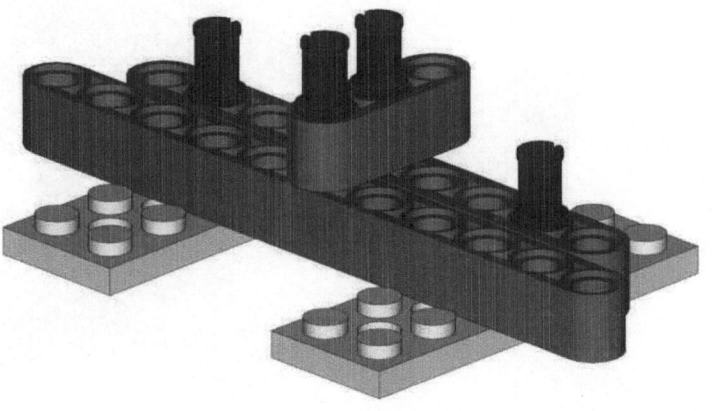

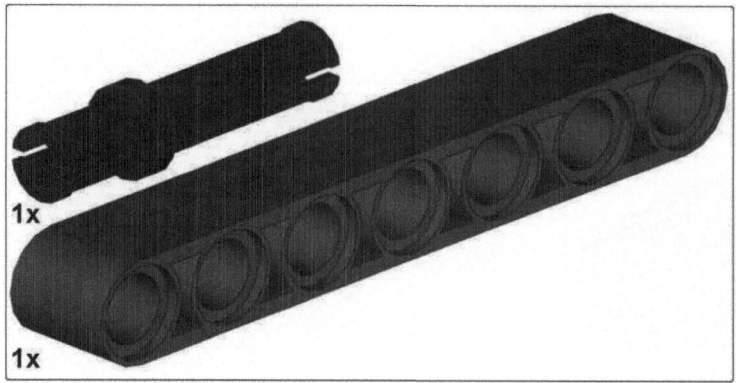

11

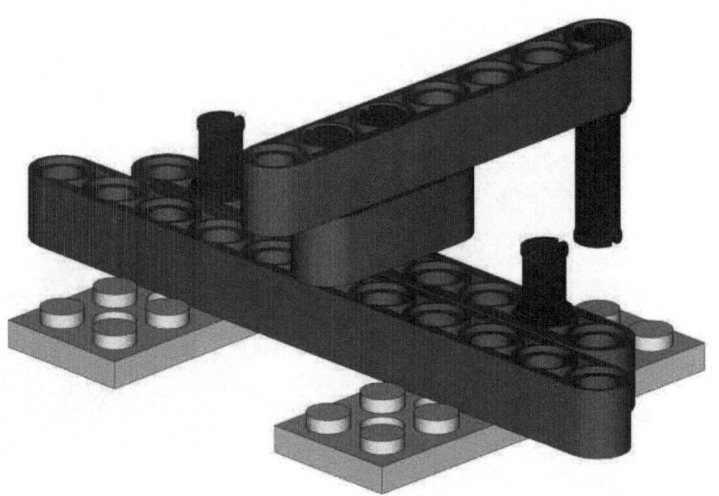

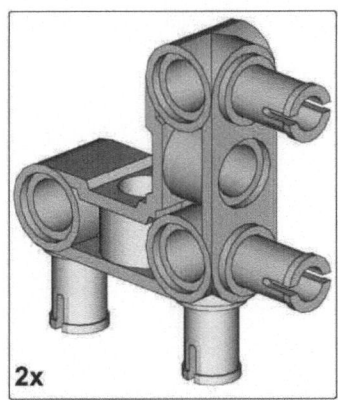

12

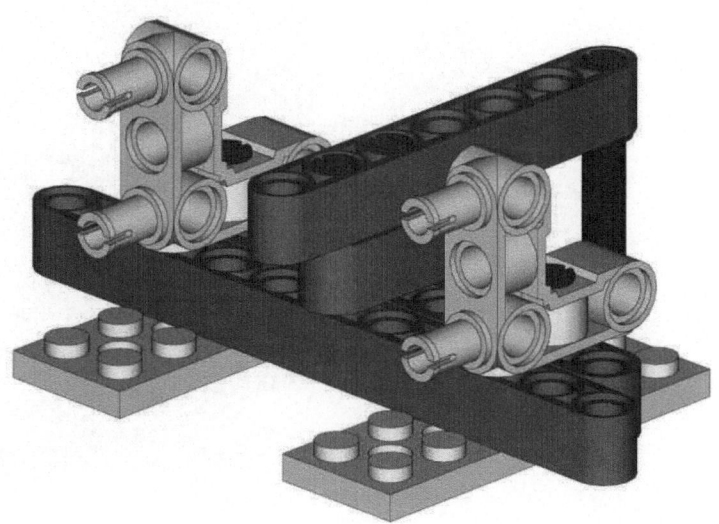

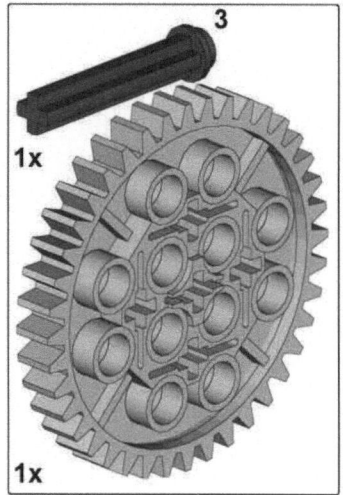

13

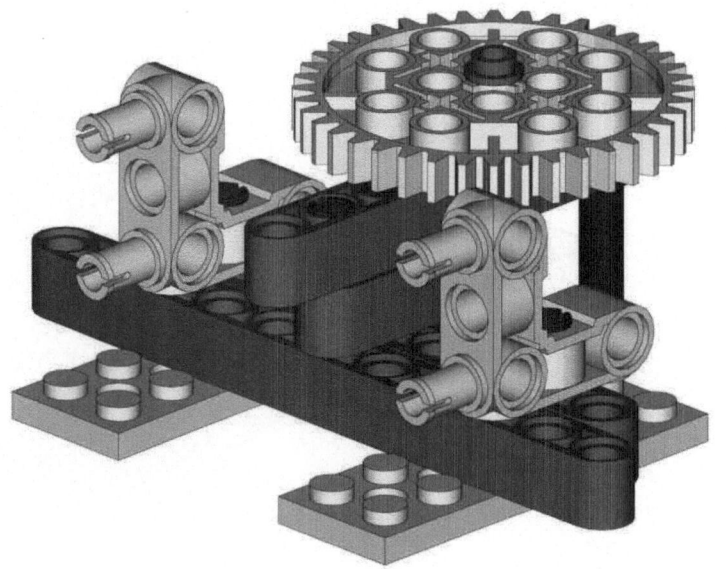

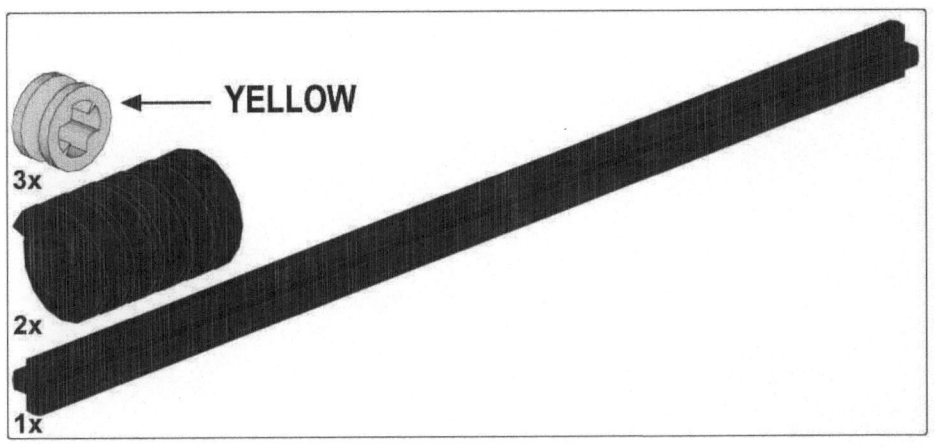

14

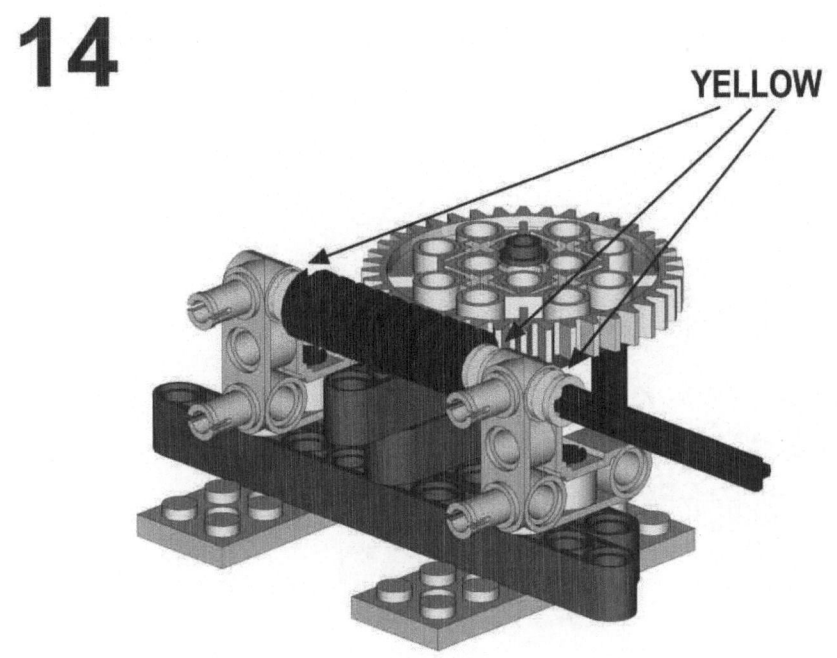

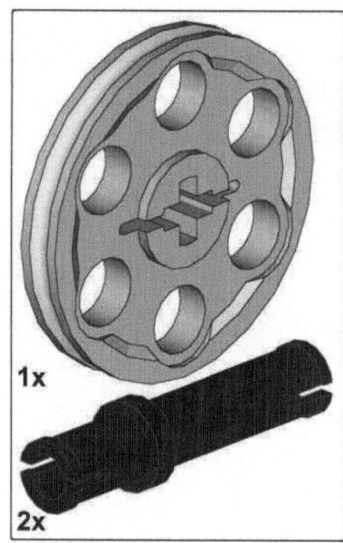

15

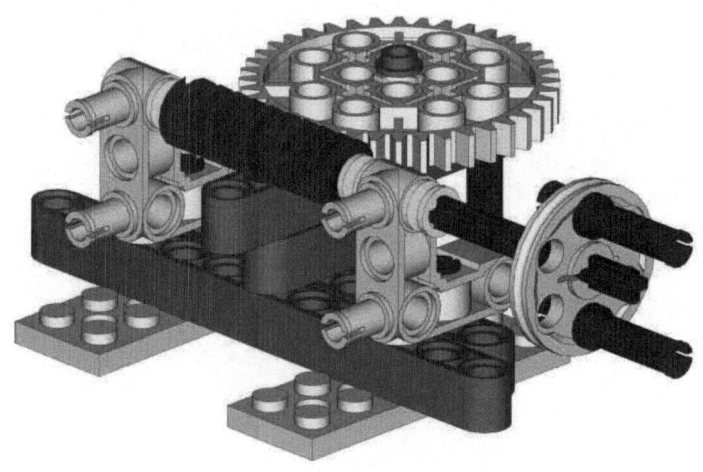

RED

2x

1x

16

RED

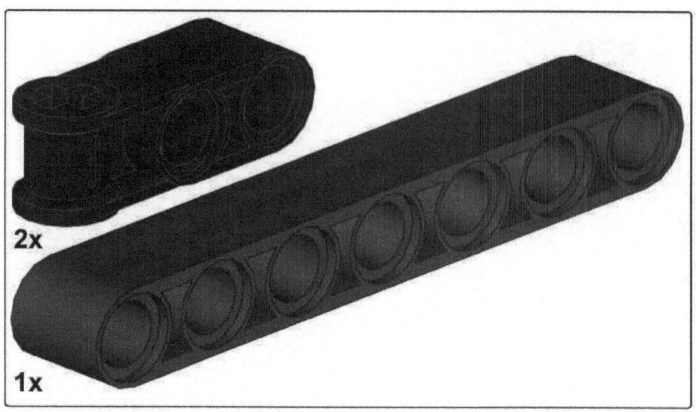

17

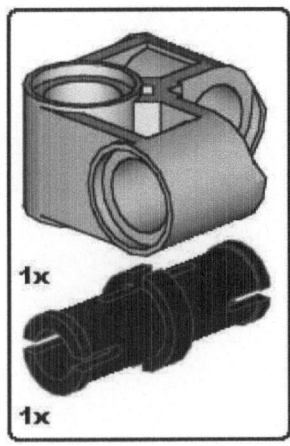

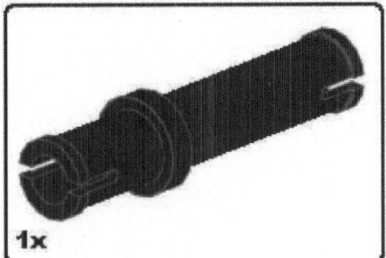

18

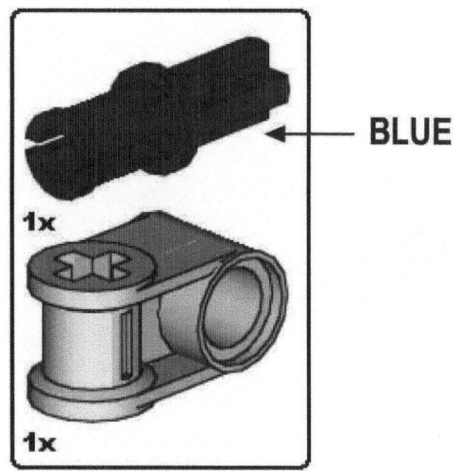

BLUE

1x

1x

19

BLUE

20

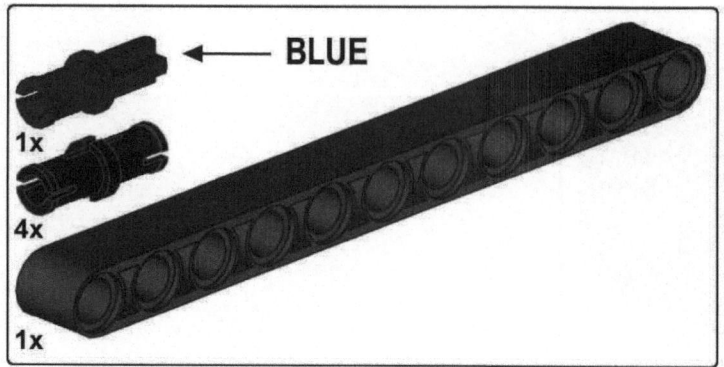

21

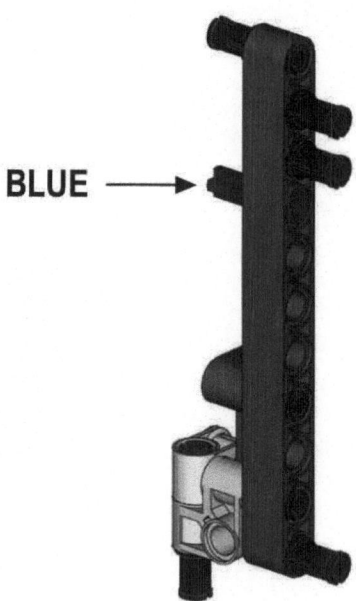

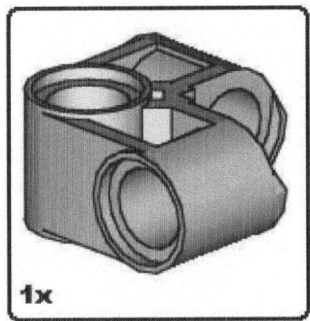

22

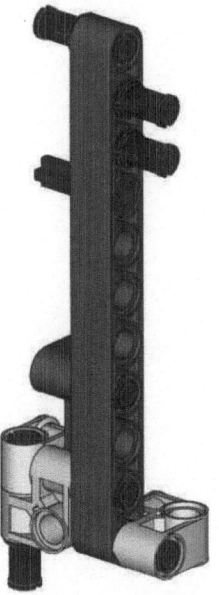

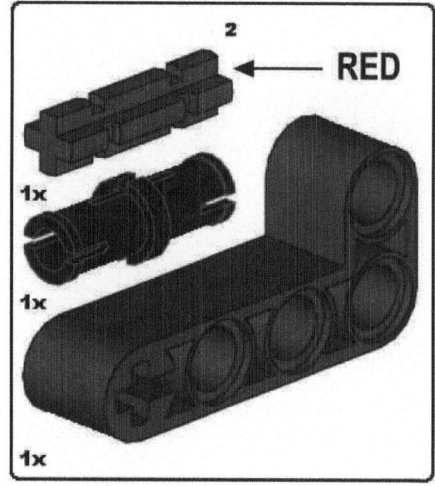

23

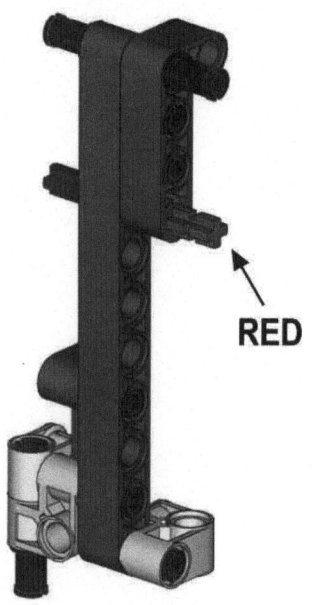

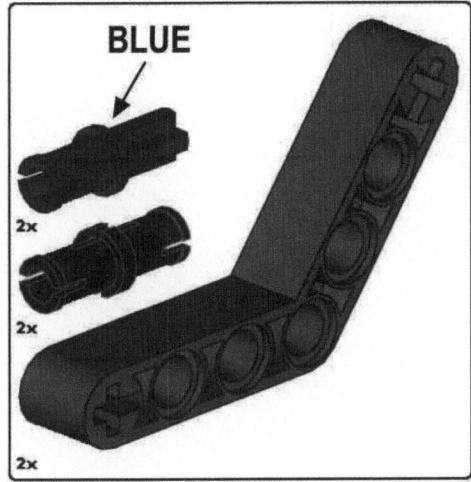

24

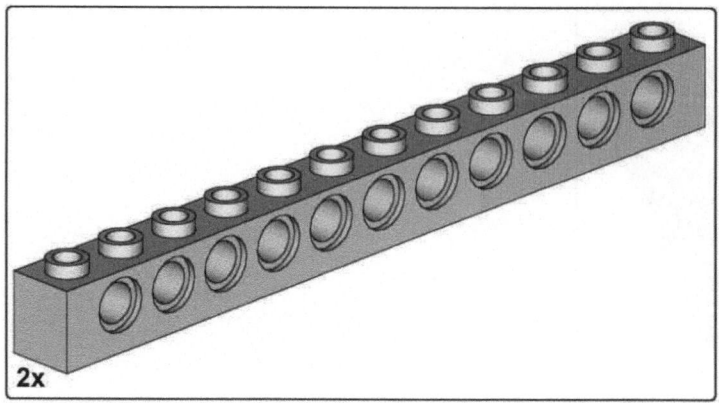

2x

25

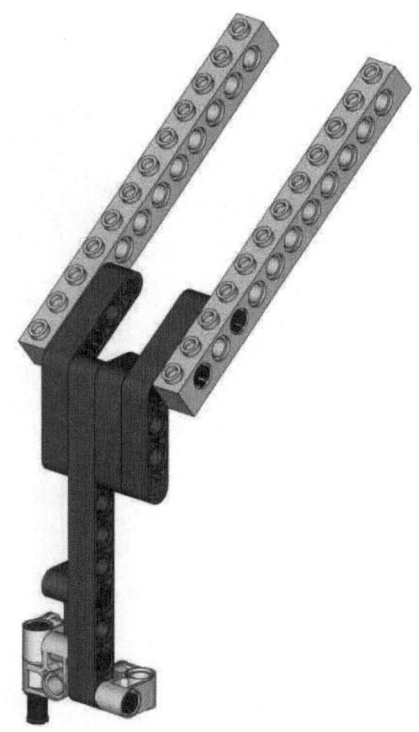

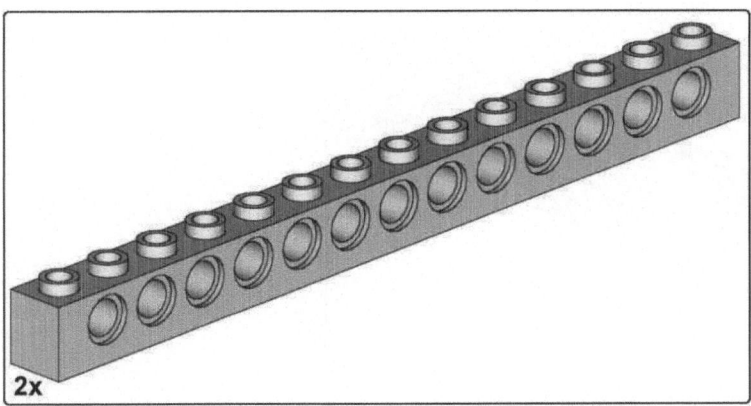

26

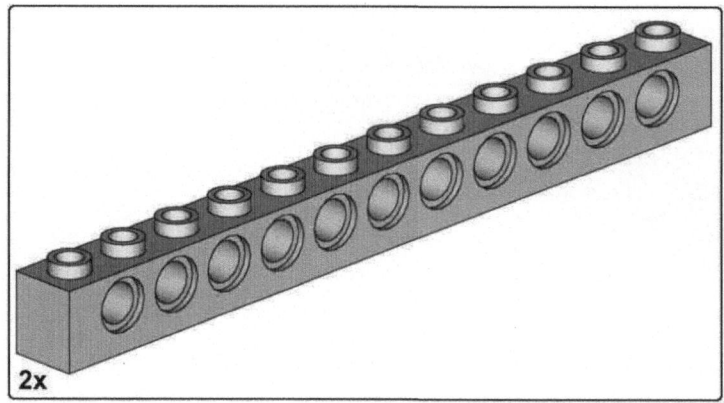

2x

27

28

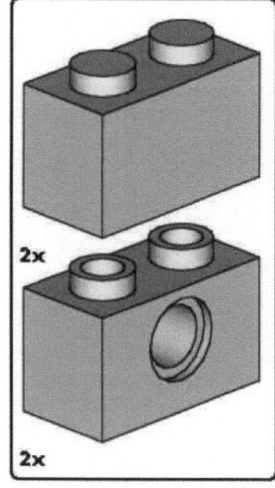

2x

2x

29

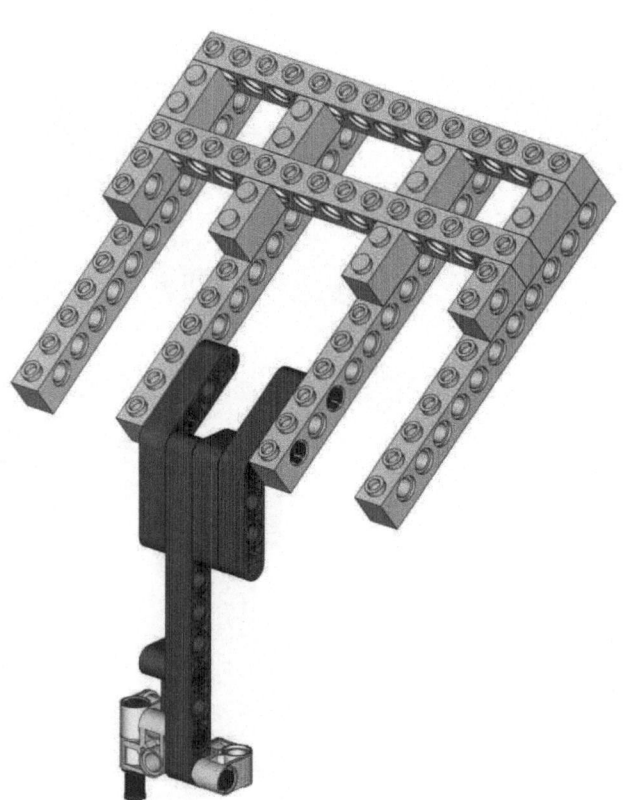

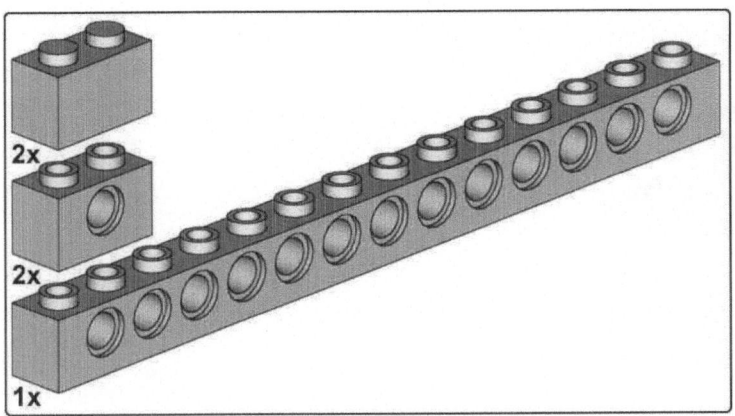

30

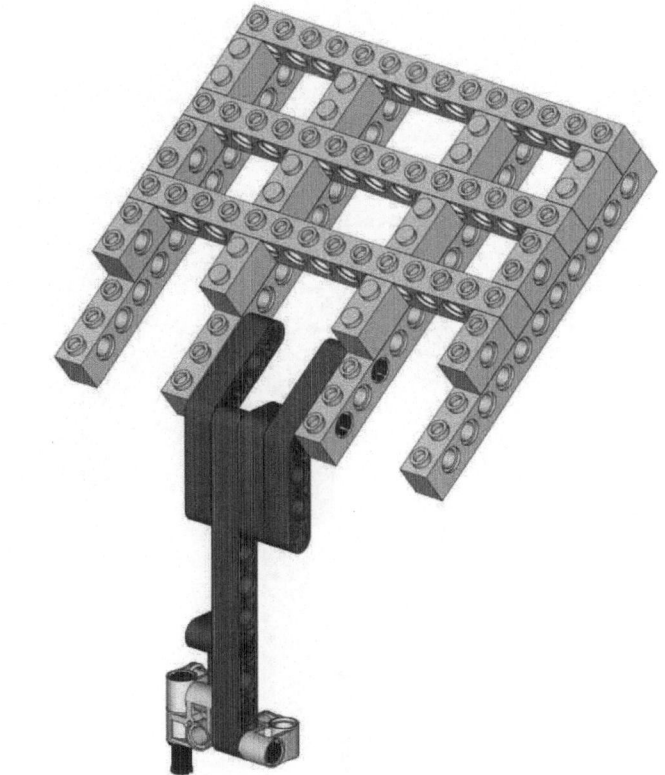

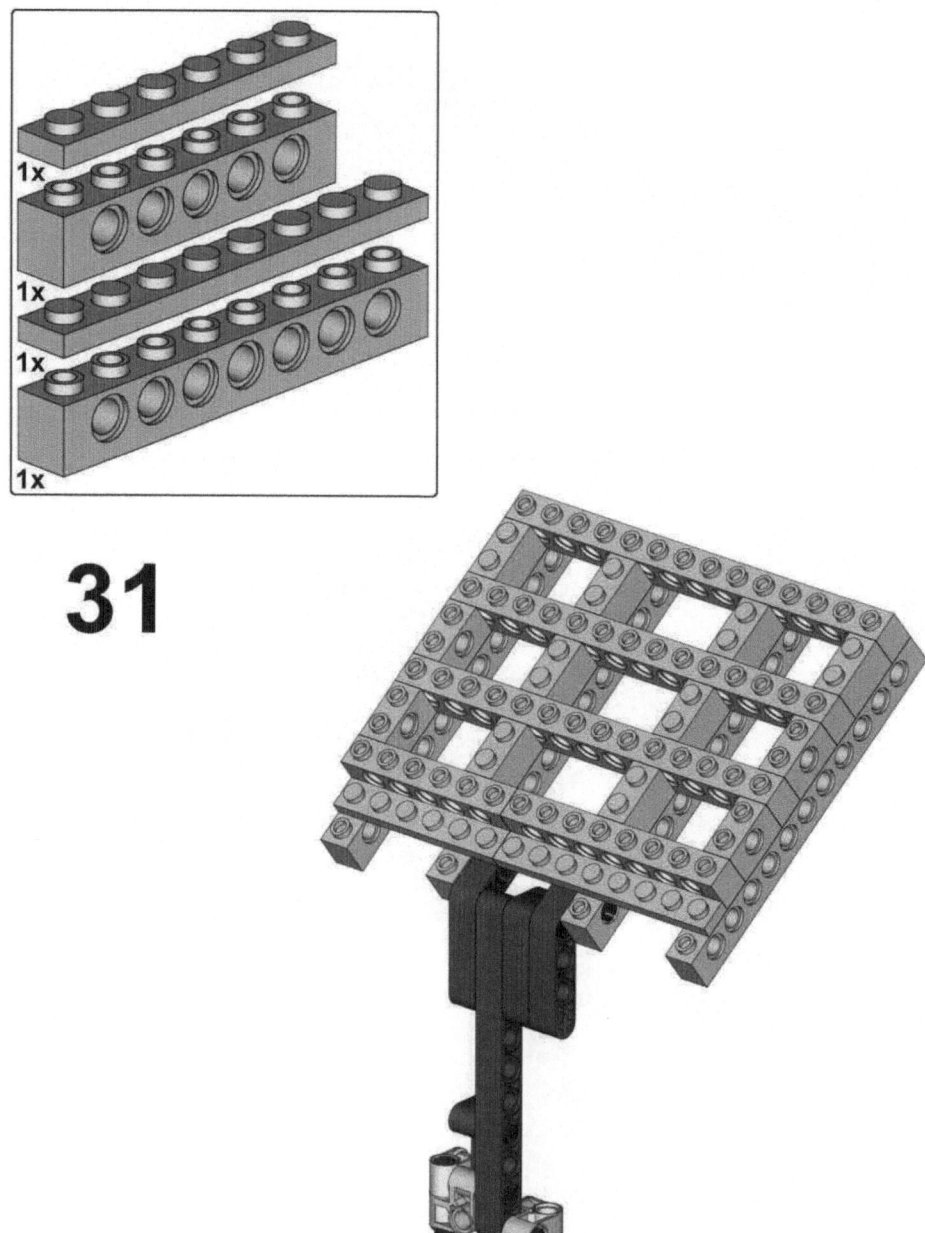

31

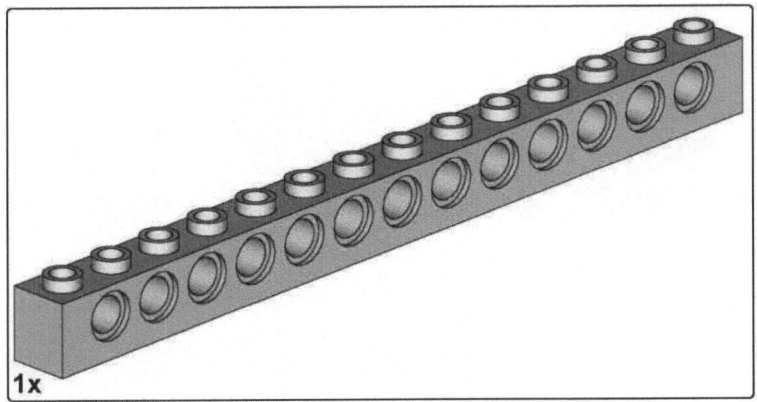

32

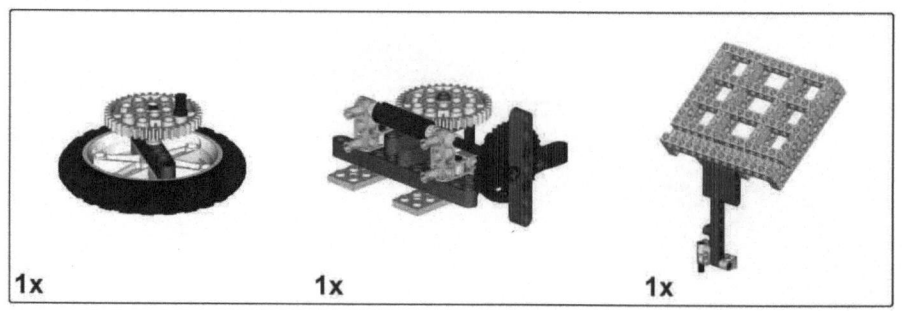

33

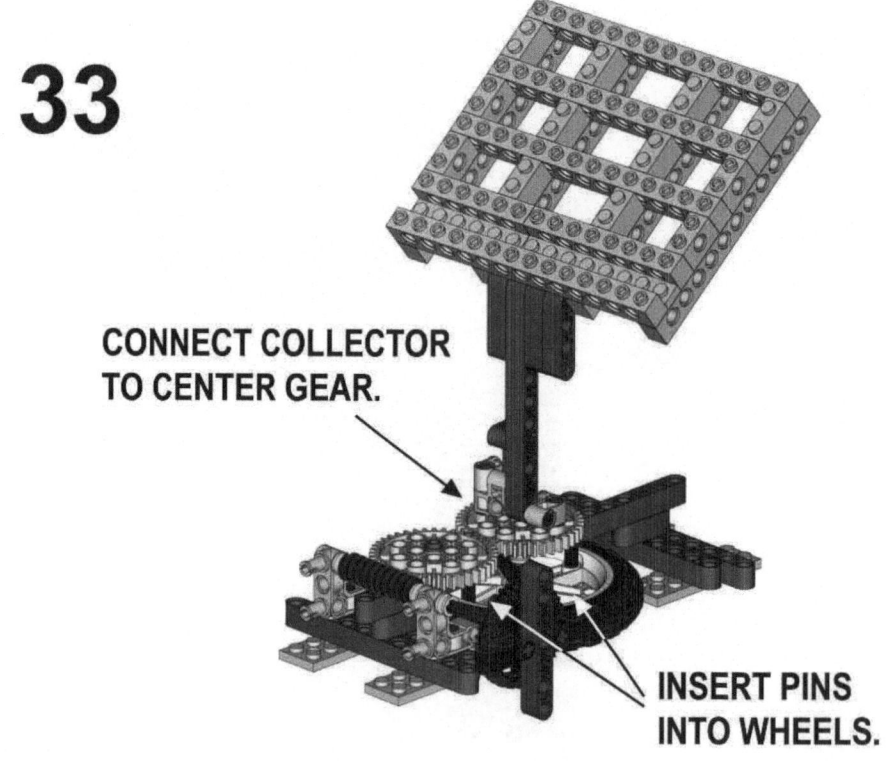

CONNECT COLLECTOR TO CENTER GEAR.

INSERT PINS INTO WHEELS.

Summary

Congratulations! You've successfully built the three models that will be needed to run the Plan B challenge. Now all that's left is to read Chapter 5 and learn how to setup the challenge so the models are placed properly. You'll also learn all about the rules of challenge, including how points are awarded. Once you've got the challenge ready to go, build and program your robot and save Mars Base Alpha!

Synopsis and Rules of the Plan B Challenge

The Mars Base Alpha: Plan B challenge focuses on your efforts as a member of Mars Base Alpha to restore power to the base after a massive outage caused by a few small asteroid impacts on the power delivery lines between the fusion plant and Alpha. You have determined that it may be possible to reroute power from some remote solar collectors by programming a base robot to redirect the power supplied by these collectors back to Alpha.

Teams or individuals must build and program a LEGO® MINDSTORMS® NXT robot to successfully interact with three models. Each model is part of the overall challenge; the combined completion of four mission objectives will determine success or failure in restoring power to Mars Base Alpha.

To successfully restore power to Mars Base Alpha, your robot will need to perform the following actions:

- Disengage the power interrupt device (PID)

- Reroute power using the power redirect (PR) to Mars Base Alpha

- Rotate the solar collector to face away from the PID (approximately 180 degrees)

- Reengage the PID

When the challenge is completed, your score will be tallied based on the successes and failures of your robot and its adherence to the challenge rules.

Setting Up the Challenge Area

The challenge area (CA), which defines the boundaries for your robot, must adhere to the requirements outlined in this section. Use Figure 5-1 as a guide for proper placement of the models, or download and print the Challenge Area Mat PDF file.

Note   You may download the full-color and black and white PDF mat files for use in this challenge by visiting www.marsbasecommand.com and clicking the Mars Base Alpha Mat button. The files may be taken to a print shop and printed as a mat (in color or black and white) to be used for the challenge area. Permission to download and use the file for private use is granted. The file may not be sold, and printed mats may not be sold. *The mat is not required to run the challenge.*

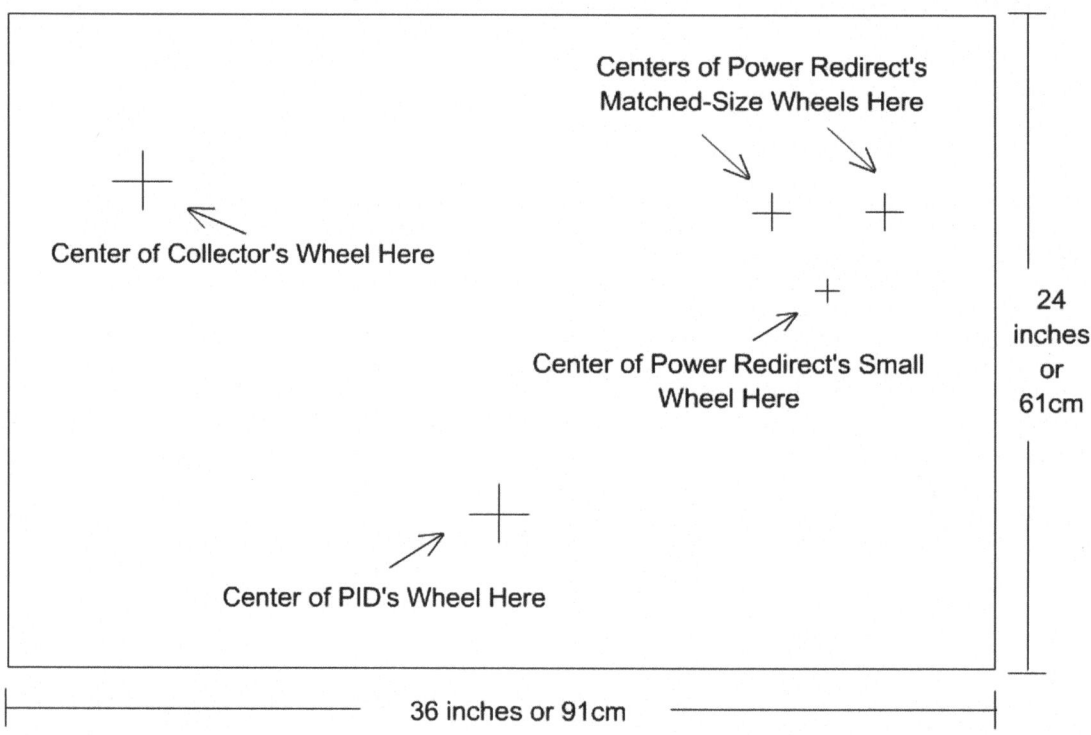

North

Figure 5-1. *The challenge area for the Mars Base Alpha missions*

The CA must be a flat surface area with dimensions of 24 × 36 inches (2 × 3 feet) or approximately 61 × 91 cenimeters (refer to Figure 5-1). The greater distance is the CA width.

The CA's boundaries can be defined using tape or other items to designate the 24" x 36" surface area.

Assign the top edge of the CA as north (for use with the challenge rules).

Place the center of the solar collector's large wheel approximately 6 inches from the top edge and 4 inches from the left edge (see Figure 5-1).

Place the center of the PID's large wheel approximately 5.5 inches from the bottom edge and 18 inches from the left edge (see Figure 5-1).

The PR has two matched-size wheels and one small wheel as its base. Place the power redirect, so the small wheel's center is approximately 6 inches from the right edge and 10.25 inches from the top edge. For the two matched-size wheels, the right wheel's center is approximately 3.75 inches from the right edge and approximately 7 inches from the top edge. The left wheel's center is approximately 8 inches from the right edge and approximately 7 inches from the top (see Figure 5-1).

Figure 5-2 shows a photograph of the models placed on the printed challenge area mat.

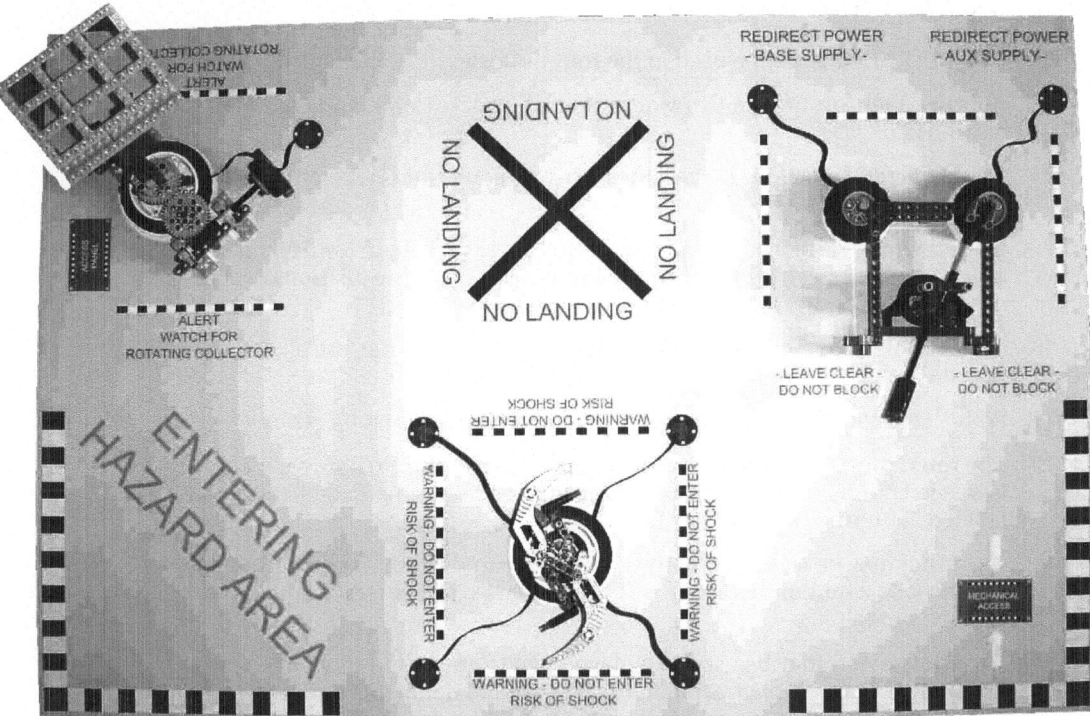

Figure 5-2. The three challenge models placed on the challenge mat

Understanding the Challenge Rules

To run the challenge properly, 16 rules must be followed. If multiple teams are running the challenge, rules can be altered or removed if all players are in agreement. The rules for the Mars Base Alpha challenge follow:

1. The robot created to attempt the missions must be built using parts from a LEGO MINDSTORMS NXT robotics kit (Retail or Education version 1.0 or higher).

2. A minimum of one robotics kit must be used to build the robot; special commendations are available for successful completion of missions using a single robotics kit.

3. A maximum of two robotics kits may be used to build the robot.

4. If a LEGO MINDSTORMS NXT Education Resource Set is used to build the mission models, any parts remaining in the Resource Set may also be used in the robot design.

5. Remaining parts from a maximum of one Resource Set may be used in the robot design.

6. Only NXT motors may be used in the robot's design.

7. Approved sensors are the Ultrasonic, Sound, Touch, Color, and Light (NXT versions only).

8. The challenge time limit is 3 minutes. The challenge ends when the time limit expires.

9. Once the robot is placed on the mat, it cannot be picked up until all mission objectives have been met or the time limit expires (see the mission objectives descriptions for success/failure descriptions).

10. If the robot is touched before all mission objectives are met, the challenge ends immediately.

11. Points are scored only for successfully completed mission objectives.

12. The robot must be placed in its starting position on the challenge area in the lower-left corner (in Figure 5-2, the starting position is defined by the area labeled "Entering Hazard Area").

13. The robot may be oriented in any direction desired (from its starting position) but must be a minimum distance of 6 inches (15 centimeters) from any mission model.

14. Tape (any color) may be used and placed anywhere on the challenge area to assist the robot with navigation. It may be used to define the challenge area boundaries as well as to provide lines to follow or points on the CA. There is no limit to the amount of tape that may be used.

15. If any portion of the robot crosses over the challenge area boundaries, a penalty is given by deducting 1 point from the total mission score *for each violation*. If the entire robot moves outside the challenge area boundaries, the challenge ends.

16. Tape, Velcro, or other items may be used to secure models to challenge area if needed as long as they do not impede the robot or make it easier for robot to complete a mission objective.

Note   Cables (for sensors and motors) that cross the boundaries will *not* be penalized for violating rule 15.

Understanding the Mission Objectives

The mission objectives can be attempted in any order but must be completed in the following order for maximum points:

> *Mission objective 1*:　　Disengage the power interrupt device.
>
> *Mission objective 2*:　　Move the power redirect from a right-pointing position to a left-pointing position.
>
> *Mission objective 3*:　　Rotate the solar collector 180 degrees to face away from the PID.
>
> *Mission objective 4*:　　Reengage the PID.

Mission Objective 1: Disengaging the Power Interrupt Device

The first objective is to disengage the PID. The following rules pertain to the placement and settings of the PID.

1. When the challenge starts, the PID must be in its engaged position (see Figure 5-3). Swivel the moving arms so that they are touching the orange pieces.

2. The robot must disengage the PID by moving the swivel arms (see Figure 5-4).

3. If the robot touches any portion of the gray claw pieces (see Figure 5-4), the challenge ends.

4. If the robot disengages the PID without touching the claw pieces, *mission objective 1 is successful*. (If robot touches the claw piece, the robot may still attempt other mission objectives, but no points will be awarded for mission objective 1.)

5. After the power redirect and solar collector missions are accomplished, the robot must return to the PID and reengage it without touching the claw pieces (See Figure 5-3).

6. If the robot reengages the PID without touching the claw pieces, *mission objective 4 is successful*. (If the robot touches the claw piece, the robot may still attempt other mission objectives, but no points will be awarded for mission objective 4 completion.)

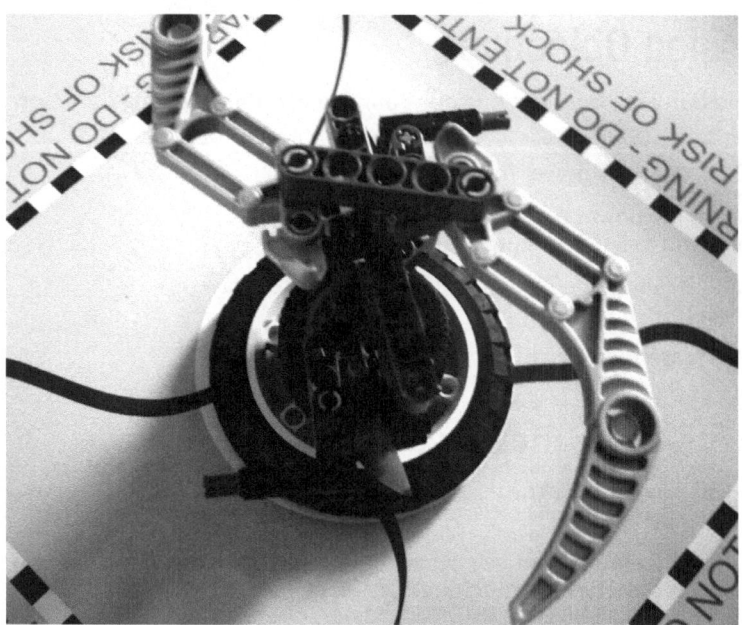

Figure 5-3. *The PID starts in the engaged position*

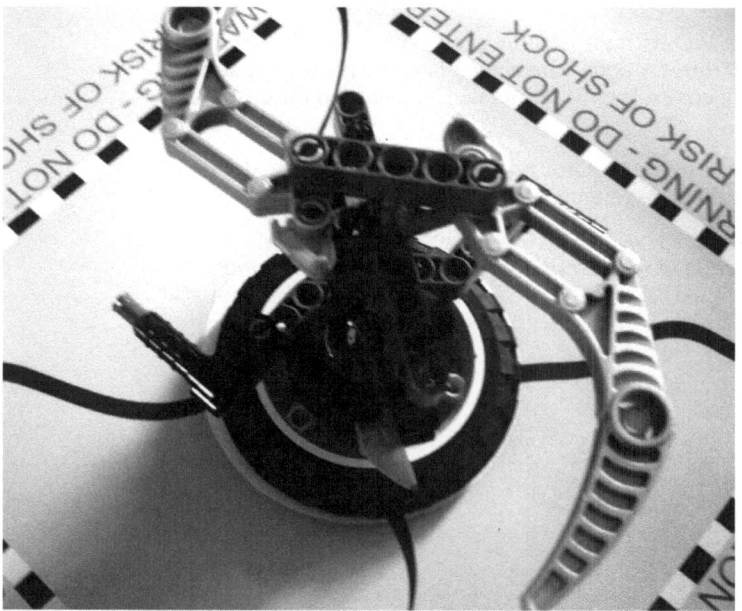

Figure 5-4. *The PID is disengaged when the swivel arms are not touching orange pads*

Mission Objective 2: Moving the Power Redirect

The second objective is to move the power redirect from a right-pointing position to a left-pointing position. The following rules pertain to the placement and settings of the PR:

1. When the challenge starts, the power redirect's swivel arm must be pointing to the right. In Figure 5-5, it is pointing at the "Redirect Power—Aux Supply" label.

2. The robot must move the power redirect's swivel arm so it is pointing to the left. In Figure 5-6, it is pointing at the "Redirect Power—Base Supply" label.

3. If the robot moves the swivel arm from a right pointing position to a left pointing position, *mission objective 2 is successful.*

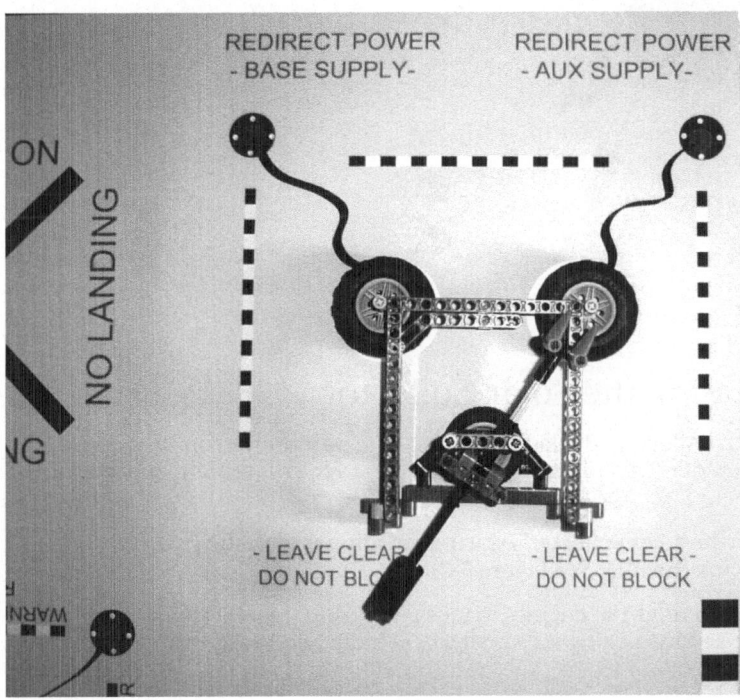

Figure 5-5. The PR's swivel arm is pointing to the right

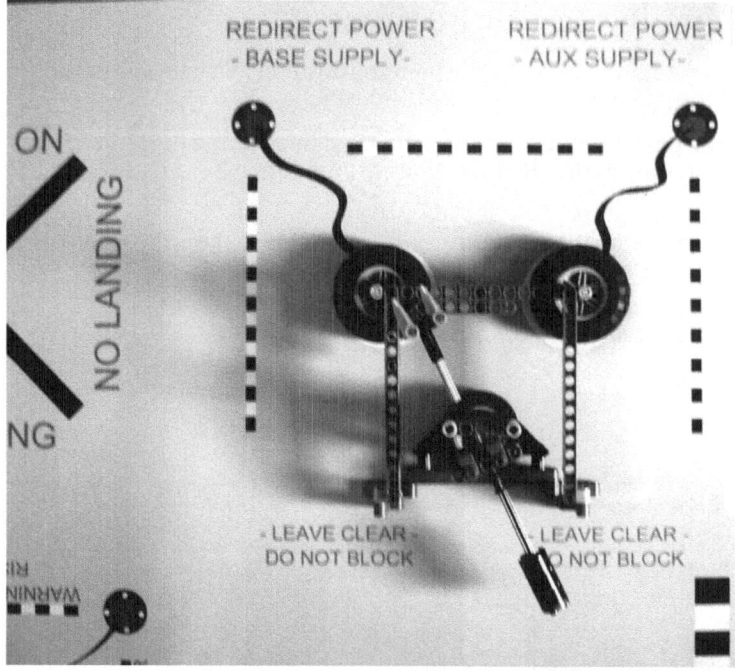

Figure 5-6. *The PR's swivel arm is pointing to the left*

Mission Objective 3: Rotating the Solar Collector

The third objective is to rotate the solar collector 180 degrees, to face the power interrupt device, by having your robot turn the spinning handle. The following rules pertain to the placement and settings of the solar collector:

1. When the challenge starts, the solar collector's starting position is with the grid portion of the collector facing the PID (See Figure 5-7).

2. The solar collector must be rotated 180 degrees to face away from the PID (see Figure 5-8). Rotation is caused by turning the small wheel indicated in Figure 5-8.

3. If the solar collector is successfully rotated 180 degrees, *mission objective 3 is successful*. The 180 degrees can be treated as an approximation; if the collector is obviously facing away from the PID, consider the mission successful.

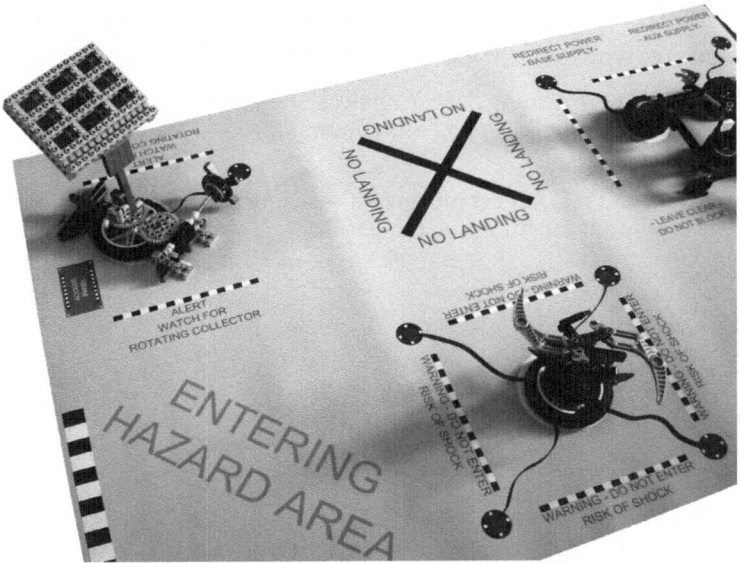

Figure 5-7. The SC is facing the power interrupt device

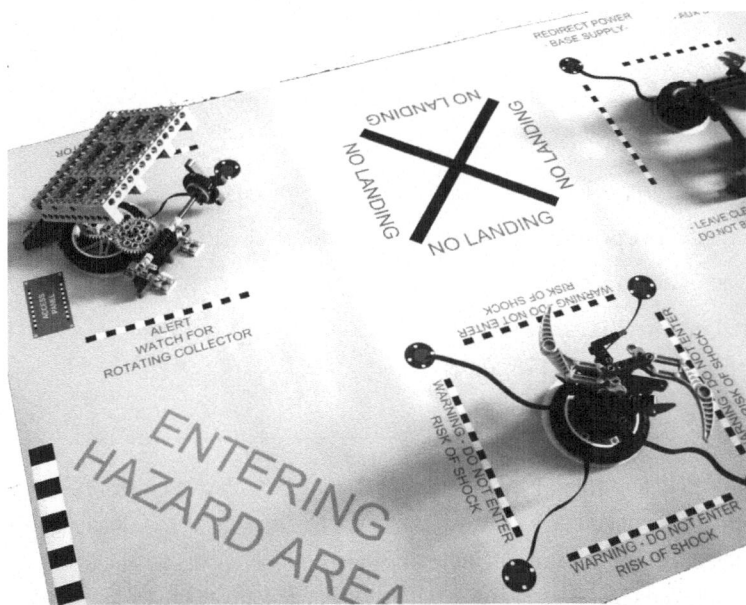

Figure 5-8. The SC is rotated 180 degrees and facing away from PID

Mission Objective 4: Reengaging the Power Interrupt Device

The fourth (and final) objective is to reengage the PID:

1. If the robot reengages the PID without touching the claw pieces, *mission objective 4 is successful*. If robot touches the claw piece, no points will be awarded for mission objective 4 completion.

2. After the PID has been reengaged, the robot must return to its starting position or leave the challenge area mat without touching any models.

Scoring the Challenge

The challenge score will be determined using the following system:

*   Mission Objective 1 completed successfully: 5 points

*   Mission Objective 2 completed successfully: 5 points

*   Mission Objective 3 completed successfully: 5 points

*   Mission Objective 4 completed successfully: 5 points

*   Any portion of Robot crosses mat boundaries: –1 point per violation

Earning Bonus Points

If you like, you can choose to assign bonus points according to the following scheme:

*   All mission objectives successful in the proper order: 5 points

*   All mission objectives successful *(in order)* in less than 1 minute: 5 points

*   Ultrasonic sensor on robot used in any manner: 1 point

*   Sound sensor on robot used in any manner: 1 point

*   Touch sensor on robot used in any manner: 1 point

*   Light sensor on robot used in any manner: 1 point

Note   Mission objectives may be attempted in any order, but bonus points will not be awarded unless the mission objectives are completed in the order described in the previous sections ("Understanding the Challenge Rules" and "Understanding the Mission Objectives"). For example, if the robot uses the Ultrasonic and Light sensors, but the robot fails to complete all mission objectives in order, the bonus points for using the sensors will not be awarded.

Using Mars Base Alpha Mission Scoring Form

Use the Mars Base Alpha Scoring Form shown in Figure 5-9 to track the completions of the mission objectives and to tally the final challenge score.

Note The Mars Base Alpha Mission Data Form and the Mars Base Alpha Scoring Form may be downloaded at www.marsbasecommand.com. Click the Mars Base Alpha Scoring button.

Mars Base Alpha Scoring Form

	Success	Fail	Points
Mission Objective 1	5	0	
Mission Objective 2	5	0	
Mission Objective 3	5	0	
Mission Objective 4	5	0	
Subtotal			◯

Any portion of robot crosses Challenge Area Boundaries

Number of Violations _____ × -1 = ◯
(multiply the # of violations by -1)

Bonus Points may only be awarded if ALL Mission Objectives are successfully completed in the proper order.

BONUS POINTS	Circle the Bonus Points if a Condition is met.
All Mission Objectives Successful In Proper Order	5
All Mission Objectives Successful in < 1 minute	5
Ultrasonic Sensor used on robot in any manner	1
Sound Sensor used on robot in any manner	1
Touch Sensor used on robot in any manner	1
Light Sensor used on robot in any manner	1

CHALLENGE SCORE
(add all values in circles) ⬭

Figure 5-9. The Mars Base Alpha Scoring Form

Frequently Asked Questions

Following are some frequently asked questions, and their answers:

1. *Do all the mission objectives have to be completed in the challenge?*
   No. You may attempt as many or as few of the mission objectives as you wish. For bonus points, however, you must accomplish all mission objectives in the proper order described in the "Understanding the Challenge Rules" and "Understanding the Mission Objectives" sections.

2. *If the robot's attachment arm, but not the robot's main body, moves over the challenge area boundaries is there a penalty?*
   Yes. Any portion of the robot—tires, attachments, sensors, etc.—that breaks the CA boundaries will be considered a violation, and a penalty will be assessed. Cables are the only exception; no penalty will result if a cable crosses the CA boundaries.

3. *Is there a height or weight limit for the robot?*
   No. However, keep the CA boundaries in mind as you design your robot; the space to move between the models is limited.

4. *What happens if a robot collides with a model and moves it?*
   If the robot successfully completes a mission objective but moves that mission's model, treat it as a success as long as no other rules have been broken. If, for example, the robot touches one of the PID's claw pieces, that mission objective fails. If the robot touches or moves a model during the course of navigating through the challenge area, you may reposition the model *only* if it will not interfere with the robot's movements. If the robot is caught on a model in such as way as to require the robot to be picked up, the Challenge ends; see challenge rule 9.

5. *What if I encounter a situation that's not covered by the challenge rules?*
   Make your own ruling on the matter that is in agreement by all parties participating in the challenge.

Adding Novice and Expert Rules

As an alternative to the standard mission challenge rules, the following novice rules should make the challenge a little easier to successfully complete:

1. The challenge time limit is 6 minutes.

2. The robot may be picked up at any time during the challenge but must begin again in the starting position (the "Entering Hazard Area" section shown in Figure 5-2).

3. A portion of the robot may cross over the challenge area boundaries with no penalty, but if the entire robot moves outside the challenge area boundaries, the challenge ends.

As an alternative to the standard mission challenge rules, the following expert rules should make the challenge a bit more difficult to successfully complete:

1. The challenge time limit is 1 minute.

2. If any portion of the robot crosses over the challenge area boundaries, the challenge ends.

3. All mission objectives must be completed in the proper order. If a mission objective is completed out of order the challenge ends immediately, and points will only be awarded for those mission objectives already completed in order.

Summary

The Mars Base Alpha challenge is supposed to be fun. While every effort has been made to keep the rules easy to follow, it's simply impossible to predict every question that will arise regarding the rules. Therefore, teams and individuals should be given the benefit of the doubt when it comes to interpreting the rules. If a rule is unclear to you, think about the overall objective of the missions, and make your best decision regarding how you will resolve or enforce that rule.

Mars Base Command is about learning, doing, and having fun—not spending large amounts of time worrying about rules. Try your best to solve the missions using the information provided, and if a robot finds itself in a situation that the rules don't address, don't worry about it! Make the best ruling you can and move on.

Finally, keep in mind that your robot doesn't have any limits on the number of times it can attempt missions. Continue to refine and test your robot until it is able to accomplish all the missions.

One last thing—if you don't like the rules, change them! Give yourself more time or allow third-party sensors or whatever will make the experience more enjoyable for you.

CHAPTER 6

Storm Front

Storm Is Coming

Mars Base Gamma, Section D, Control Center
September 25, 2062 at 3:12 AM (Greenwich Mean Time)

"We only have how long?" asked Specialist Rick Webber.

Commander Colton Dennis turned and smiled. "I keep forgetting this is your first Martian storm, isn't it?"

Rick nodded.

"We've got about four hours. Plenty of time, Rick. This is a category 3 storm, and we've been through dozens of category 4 storms with no real problems. It's just standard procedure," said the commander.

Rick was sitting at a console, looking at the satellite imagery of the approaching storm. "Do the storms have names?" he asked.

"No," replied Dennis. "But some of the base personnel have suggested it." The Commander stood up from his chair and stretched. "Don't worry. Gamma is built well and can handle this. It's your landers that I'm worried about. How many do you have out there?"

Rick typed a few quick commands and reviewed the data on the screen. "Eight right now, and one of them is really pushing the limits on distance. Even at full speed, I'm not sure."

The commander walked over to Rick's station and looked at the screen. "Which one is that?" he asked, pointing at the screen. "PL372?"

"That's Grumpy," replied Rick. "I wish it was Speedy."

Commander Dennis nodded. "Well, if Grumpy doesn't outrun that storm, he's going to be a lot grumpier. I'm going to start my inspection. Back in a bit."

Rick frowned. His landers were rugged little machines, but he wasn't certain they could survive a meeting with a storm. He typed a few more commands and got a status report from all his landers.

Landers Bashful, Speedy, Buzz, and Scrapes were less than one hour's travel time away. No worries there. The terrain between those four landers and Gamma was fairly even; they'd be able to navigate into the base garage with no problems.

Squeak, Bumper, and Little Gear were no more than three hours away at normal speed. Little Gear, the second-to-farthest away, had successfully planted his collection of sensors and could move quickly. *Come on home, guys*, thought Rick.

Rick's eyes drifted to the satellite image and then to his landers' GPS positions. He focused on PL372. "Grumpy," he whispered. "Move it, buddy. Storm is coming…"

Lockdown

Mars Base Gamma, Section D, Control Center
September 25, 2062 at 4:47 AM (Greenwich Mean Time)

Rick pressed the Send button for the radio. "Marcus, can you verify the status of the landers, please?"

A few seconds passed before the radio squawked. "Bashful and Speedy are in their recharge pens. I can see Scrapes through the porthole," replied Marcus Kennedy, "No sign of Buzz yet."

"Thanks," replied Rick, releasing the button.

"We've lost landers to storms before," said Major Steven Landis from across the room. "I know they're just machines, but it still feels like losing a team member."

Rick sighed. "After you spend hours repairing, upgrading, and maintaining them, you do get attached. I think that's why I name them," replied Rick. "Every one of them has a personality."

Major Landis nodded. "When I arrived at Gamma six months ago, we had a storm pop up quickly. We had only about an hour to prepare. I lost an Explorer bot that I'd sent out to do a repair on a power conduit. I was able to retrieve some of the parts," he said slowly. "Let me know if there's anything my team or I can do to help."

"Right now," said Rick, "everything depends on Grumpy being able to quickly navigate around any obstacles. He's too far away for us to retrieve by driving out there in a rover."

"We're in lockdown," said Landis. "Even if he was in range, no one can leave the base now. Hopefully, he won't get into any tight spaces."

"Well, Grumpy does have one advantage over the other landers," said Rick. "He's got one of those experimental grappling hook rail guns. If he gets stuck, he can fire the grappling hook and pull himself out."

"Let's just hope it doesn't come to that," said Landis, reviewing a computer screen. "Estimates show the storm will hit us in one hour, fifty-two minutes. Where is Grumpy?"

Rick typed a few keystrokes and frowned. "Unless there's a major reduction in the storm's formation and speed, Grumpy should be about three or four kilometers from the base when the full force of the storm hits."

Good Luck, Boys

Mars Base Gamma, Section D, Control Center
September 25, 2062 at 6:57 AM (Greenwich Mean Time)

Marcus Kennedy closed the inner airlock door and secured it. He then popped the locks on his EV suit's helmet, removed it, and took a deep breath. Behind him, seven landers sat charging in their stations.

Marcus pressed a button on the wall radio. "Rick, Squeak and Bumper just arrived. No sign of Grumpy. I waited as long as I could, but I had to lockdown the garage. Sorry."

"Understood, Marcus. Thanks," replied Rick. "We'll monitor him from here."

Marcus looked out the porthole. He couldn't see more than a few meters outside the base. The storm was kicking up the soil, and he could hear small pebbles popping against the outer wall.

Rick's landers were rugged little machines, but Marcus seriously doubted they would survive a category 3 storm. In the past two years, Marcus's team had never lost one of its base maintenance bots to a storm. The maintenance bots were designed to operate in extreme conditions and only came inside the base for basic checkups and battery swapouts once a month.

Marcus typed a few commands on his terminal and checked the status of his fifteen maintenance bots. On the screen, fifteen small green checks appeared, one next to each robot. Each robot was settling in for the storm; attachment tools had been pulled inside each robot's body, sensors covered inside protective shells, and bodies snuggled up against the base walls to reduce the chance of being knocked over by debris.

Marcus smiled. "Good luck, boys," he said, turning off the light to the garage. Marcus grabbed his helmet, closed the door to the garage, and left to go grab some food.

Damage Detected

Mars Base Gamma, Section D, Control Center
September 25, 2062 at 7:25 AM (Greenwich Mean Time)

Grumpy's sensors had no protection against the strong winds. The small lander's processors were overloaded with sensory input, and the lander could not respond quickly enough to the changing conditions. The last GPS reading before the storm prevented any further calculations put Grumpy at approximately three kilometers from Gamma. Once GPS failed, Grumpy did what it had been programmed to do; drive in as straight a line as possible and use obstacle and rotation sensors to stay on track.

But it wasn't enough. Grumpy's obstacle detection abilities were overwhelmed by the wind and changing surface. Grumpy was unable to detect the sudden change in surface angle and tipped forward into the small crater. As it tumbled, Grumpy impacted with a few medium-sized rocks. Damage was detected.

Grumpy's self-preservation protocols were activated. All non-essential systems were powered down. Saving battery power was a priority. The high-capacity solar panel was of no use in a storm like this, so Grumpy continued to follow his programming; stop moving, turn on the emergency beacon, and wait for help.

Worth a Shot

Mars Base Gamma, Section D, Control Center
September 25, 2062 at 8:42 AM (Greenwich Mean Time)

"I understand, Rick, but the lockdown is in place for a reason," said Dennis as he signed reports and checked screens. The Control Center was crowded with personnel, but Rick was weaving through the tight quarters, following the commander from station to station.

"He's less than three kilometers away. I could get to him in less than ten minutes in one of the rovers," said Rick. "He's only got about an hour's worth of battery power left."

"The storm's winds won't subside for at least another eight hours," said Dennis. "Command won't authorize anyone going outside until the wind speeds reduce substantially."

Marcus indicated a line on his screen, and Dennis signed it. The garage had received no damage, and all garage technicians were checked in. Marcus gave Rick a brief smile. "We'll go get him as soon as the winds die down."

Rick shook his head. "The last bit of data I got from Grumpy seems to indicate he's fallen into a crater. If the winds don't stop, he'll be covered by soil. Once his battery dies, the beacon stops and it's going to be very difficult to find him. All his rock samples and mapping data will be lost."

Commander Dennis sighed. "Believe me, Rick. I wish there was something we could do. I'm open to ideas, but I'm not allowing any of my people to go outside until I get approval from Command."

119

Marcus rubbed his chin. "No people, huh? What about a nonhuman rescue team?"

Dennis and Rick both looked at Marcus.

Marcus pulled up a schematic on his computer screen. "Look, my maintenance robots are still outside. They've not got the greatest processors, but they're strong and they're powered up. Why can't I send one of them to retrieve Grumpy?"

"Can you remotely control them?" asked Rick. "Grumpy may have some damage or need some help to get out of the crater."

"Remote control won't be possible with the weather conditions the way they are, but each of the robots is programmed with some standard retrieval and repair protocols. I can't guarantee anything, but shouldn't we at least try?"

Rick looked at Dennis. "Commander? You won't be putting anyone at risk, and the maintenance bots are already outside."

Dennis looked at the computer screen. "It's worth a shot, Marcus. I've got a hundred things to do over the next few hours, so can the two of you handle this rescue operation on your own?"

Rick smiled. "All we've got to do is get Grumpy out of the crater. Even if his battery dies, his data and samples will be secure and we'll still be able to locate and retrieve him later."

"Alright. Do it," said Dennis. "I'll check in with you both later. Good luck."

CHAPTER 7

■ ■ ■

Rescuing the Lander Mark VII

To successfully complete the Mars Base Gamma challenge, you will be required to construct a robot that will interact with the damaged Lander Mark VII. This model will be built using parts from the LEGO Education Resource Set.

The fully assembled lander is shown in Figure 7-1; note that the challenge will involve performing repairs to the lander and that various subassemblies that make up the lander will be removed to set up the challenge (described in Chapter 9).

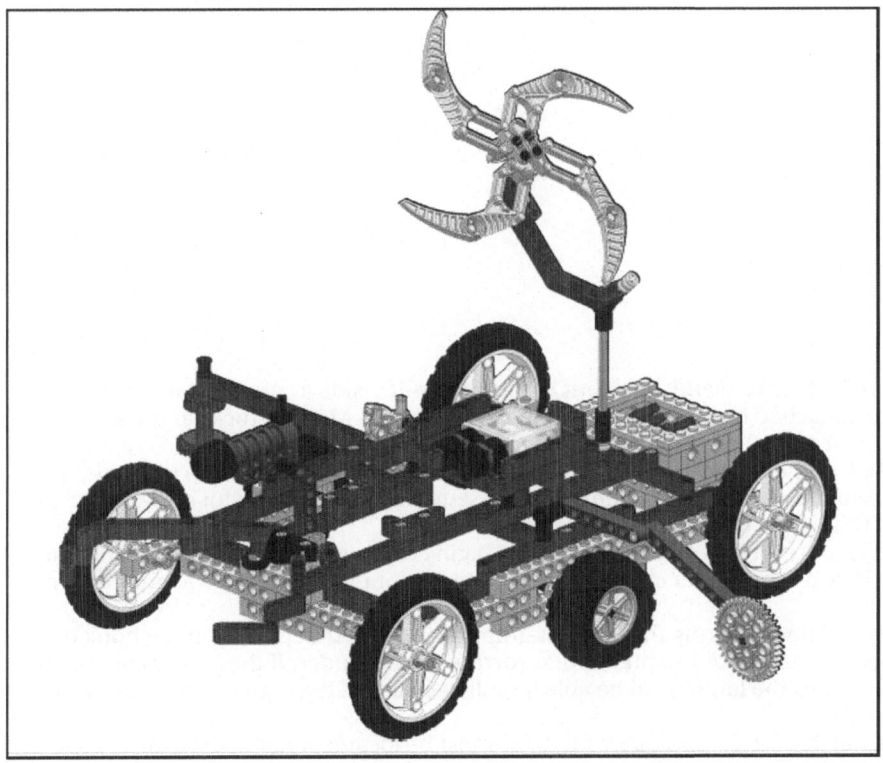

Figure 7-1. The fully assembled Mars Lander Mark VII

The lander has a handful of moving parts in addition to subassemblies that can be removed. Other parts (such as the collected Martian rocks) are pieces that will not attach to the actual lander but instead collected in a small tray or box.

Chapter 9 will give you the details for the proper placement of the damaged lander and its various subassemblies. Chapter 9 also provides details on the challenge area; once again, you can quickly assemble the challenge area with nothing but the lander, its various subassemblies, and some tape to define the boundaries of the playing field. As with the Mars Base Alpha challenge, you can also download a mat that can be printed in color or black and white and used as the challenge area.

Tackling the Lander Challenge

Your first task will be to use the building instructions at the end of this chapter to build the completed Lander Mark VII. Chapter 9 will provide details on how to "damage" the lander in preparation for running the Mars Base Gamma challenge. For now, however, just build the lander and attach all parts as described in Chapters 7 and 8.

After you've assembled the lander, you'll notice that the entire assembly consists of smaller subassemblies as follows:

- Chassis

- Battery

- Cutter

- Front arms

- Grappling gun

- Samples box

- Communications umbrella

- Vacuum

The lander has survived a Martian sandstorm but has found itself inside a small crater. Various components have become detached from the lander's main body and its Martian rock samples (collected from a variety of locations) have become scattered.

The chassis is fairly self-explanatory; it is the largest part of the lander and makes up the main body to which the other subassemblies attach. In addition to subassemblies, a small handful of "Martian rocks" will also be part of this challenge.

The majority of the challenge will involve your robot salvaging damaged pieces from the lander and either placing them anywhere on the lander chassis body for storage or in their original locations (for bigger bonus points).

Additional challenges will involve firing the lander's grappling gun so that the grappling hook on the end lands outside of a specified area (see Chapter 9) that surrounds the lander. If the lander can fire its grappling hook outside this area, the lander will be able to pull itself out of the crater on its own battery power.

To fire the grappling gun, however, the lander's battery must be located and placed in a very specific location on the chassis. If the battery is unable to be properly placed, there is still one more opportunity to save the lander; your robot must hook up a tow line to the lander's front hook and pull the lander out of the crater.

If the lander is able to be removed from the crater, either on its own power (firing the grappling gun) or by being pulled out (by your robot), additional points will be awarded based on the number of damaged subassemblies that were either placed on the lander chassis body or reattached (for bigger bonus points) to the lander.

There are no challenge area boundaries to be concerned with; your robot can roam as far or as near to the lander as you see fit to let it. Your robot may even touch the lander with no penalty points.

The two biggest challenges you're likely to face are the proper placement of the battery back in the lander and the firing of the grappling gun.

The most difficult bonus points that can possibly be awarded are related to the collection of the Martian rocks scattered in the crater; these are small pieces and can be difficult to pick up.

All remaining subassemblies will need to be picked up by your robot (using any method you can come up with that stays within the rules defined in Chapter 9) and placed either on the chassis body itself or carried out of the crater by your robot. The key word is "carried," because pushing the subassemblies out of the crater will damage them and result in no points.

Examining the Challenge Area

After you've successfully built the Lander Mark VII, you'll want to examine it carefully, noting how the various subassemblies connect to the lander (if you wish to aim for the larger bonus points awarded for original placement back on the lander).

Once you've examined the lander properly, you'll next need to consult the challenge rules in Chapter 9 regarding the proper placement of the various "damaged" subassemblies.

As was suggested for the Mars Base Alpha Challenge, one good method for approaching this challenge is to attempt to build and program a robot that can successfully find, pick up, and place the battery in its proper location. The largest number of points is awarded if the lander can successfully pull itself out of the crater using its grappling gun. So, make finding and firing the grappling gun your next highest priority.

If you are unable to build a robot that can pick up and place the battery, you can still attempt to complete the challenge by building a robot that can hook on to the lander's front hook and pull it out of the crater area. (The crater area is going to be defined as a circle of diameter X inches (Y centimeters) with the chassis placed directly in the center of the circle.)

Once you've successfully built a robot that can either replace the lander's battery and fire the grappling gun or latch onto the lander's front hook and pull it out, you'll next want to try and grab as many bonus points as possible by collecting the various damaged subassemblies scattered around the challenge area and either carrying them out of the crater (with your own robot) or by placing them on the lander's chassis body before the lander leaves the crater.

An extra motor or two will be beneficial when it comes time to lift and move various subassemblies, such as the battery or communication umbrella. You will want to experiment with lifting, hooking and reeling, and other techniques for moving objects.

Given the time limit on the challenge, it is unlikely that you'll be able to design a robot that can rescue all damaged subassemblies, replace the battery, fire the grappling gun, and collect all the Martian rocks. Determine the various parts of the challenge that you and your robot can successfully accomplish in the time limit that also maximize the possible points awarded. Be conservative at first, and if you achieve success with a portion of the challenge, move on to adding more rescued subassemblies or more Martian rock collecting to future robot runs.

Building the First Half of the Lander Chassis

The first half of the building instructions for the lander are provided in the remainder of this chapter. Each image shows the pieces you will need to locate in the LEGO Education Resource Set and their quantities. You will also see where these pieces are to be placed.

Chapter 8 will continue with the building instructions for the remaining sub-assemblies. When done, you'll have a fully-built Lander that looks like the one in Figure 7-2.

Figure 7-2. The actual Mars Lander Mark VII built from LEGO Resource Set parts

Note If you are uncertain about the placement of a piece in a figure, jump ahead to the next image or even go back to a previous image to make certain you've built the model correctly so far. Examine the figures carefully and you should be able to correctly identify the pieces and their final location on the model.

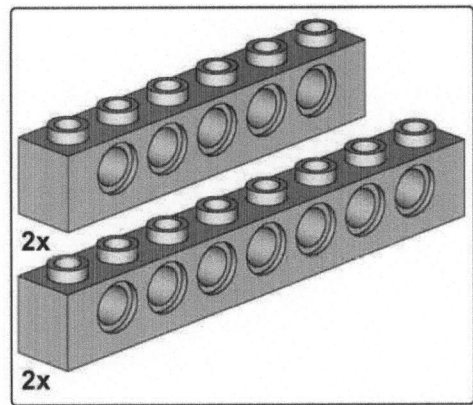

1

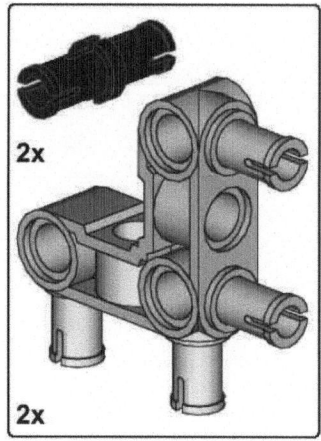

2

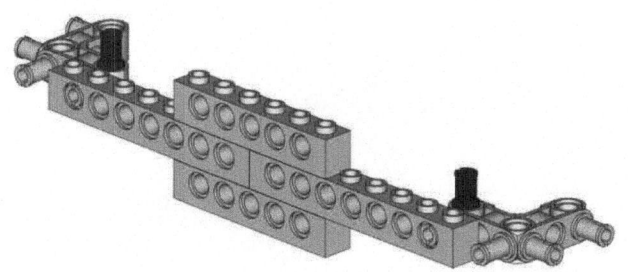

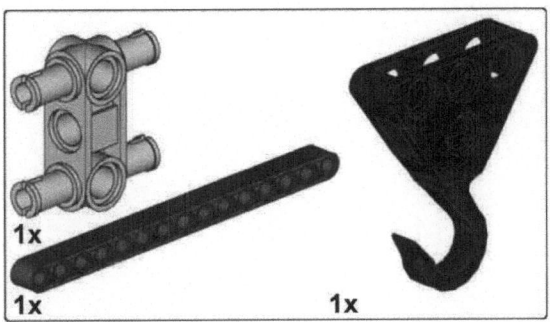

3

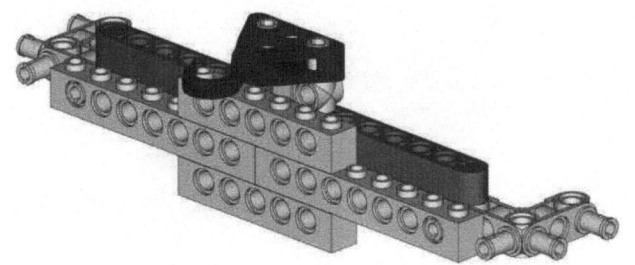

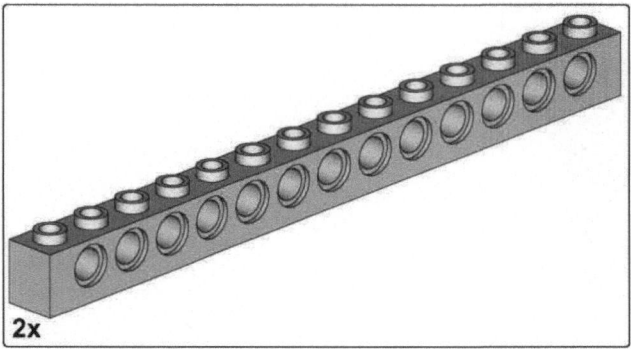

4

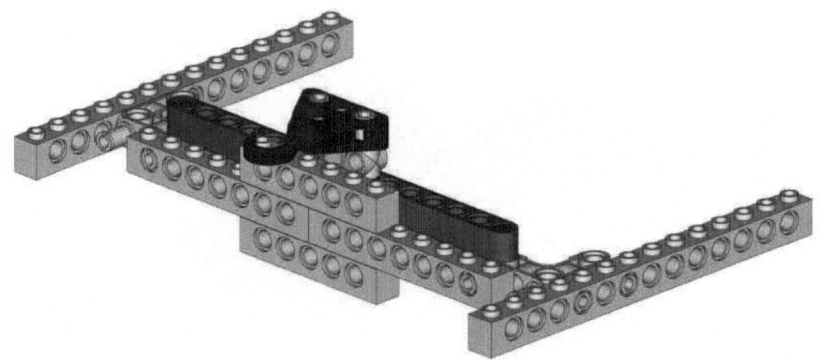

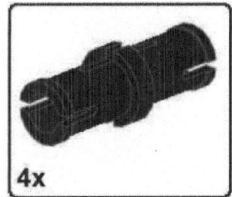

4x

5

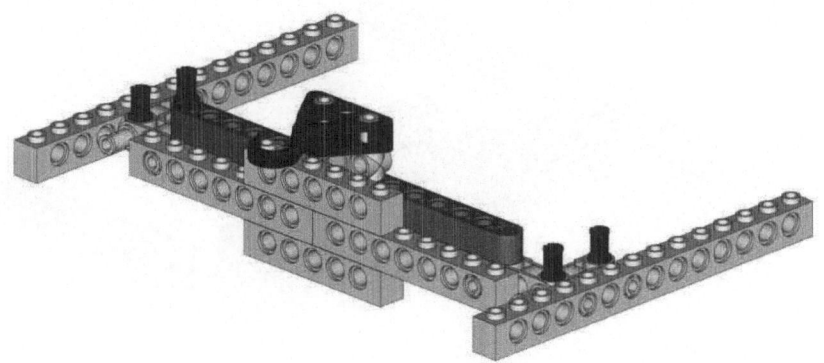

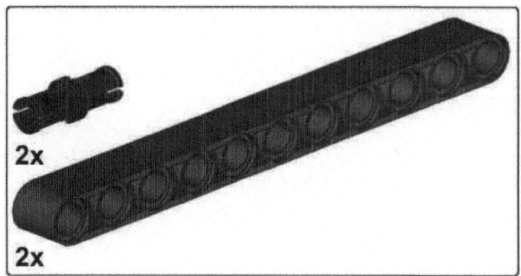

6

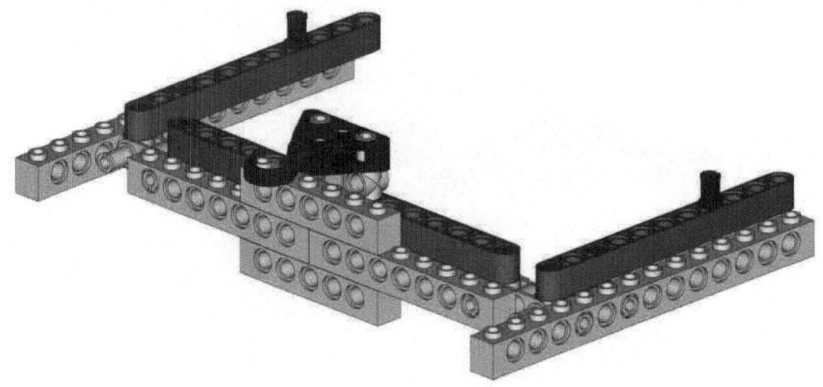

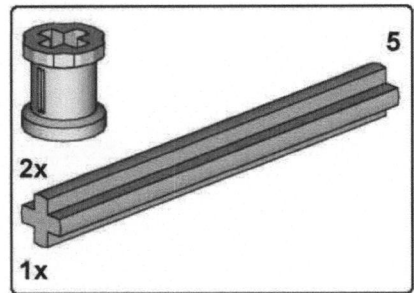

7

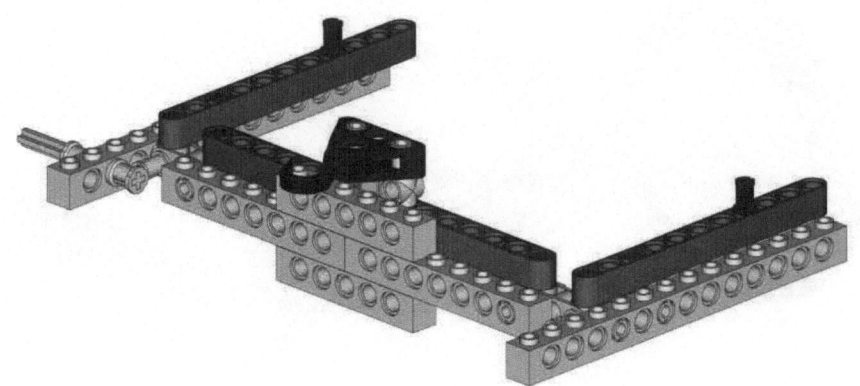

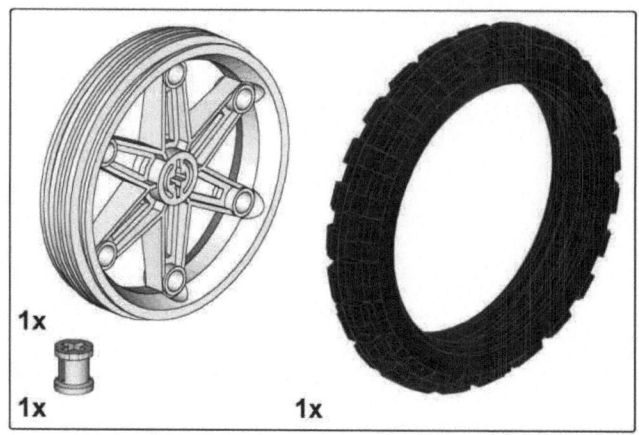

1x

1x　　　　**1x**

8

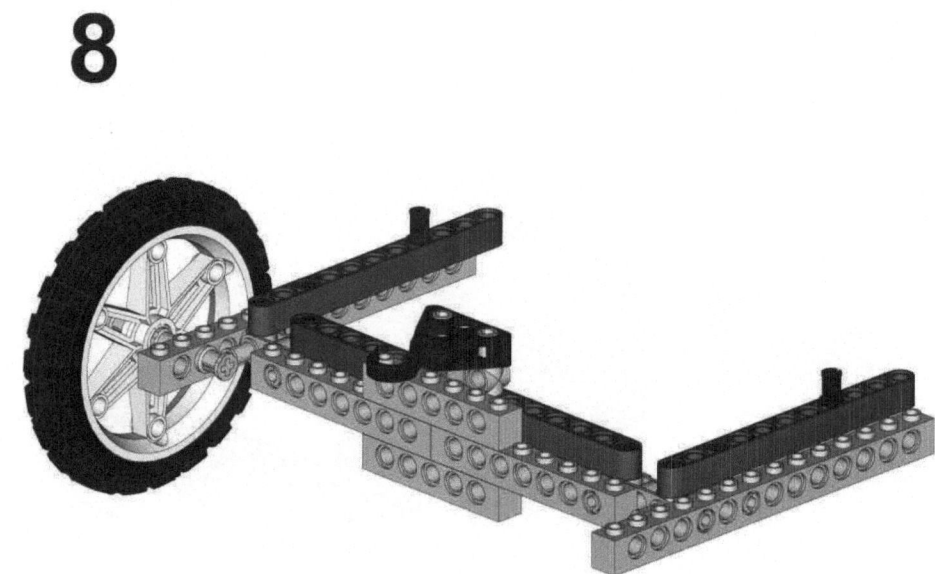

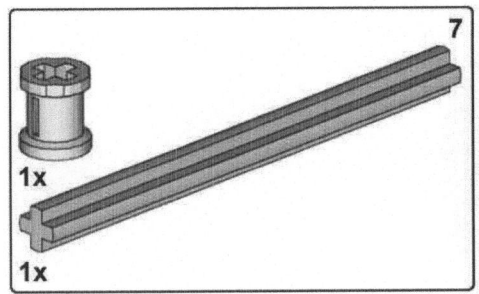

9

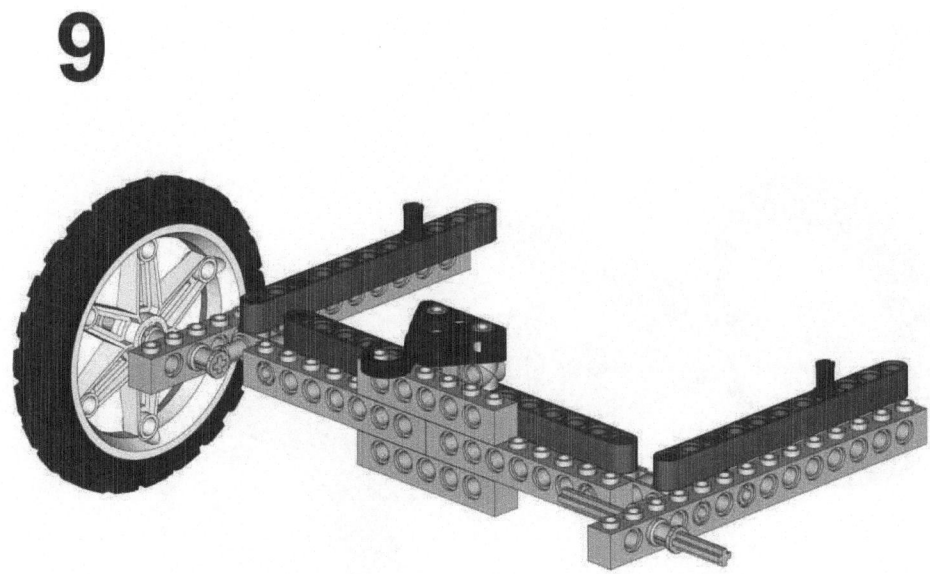

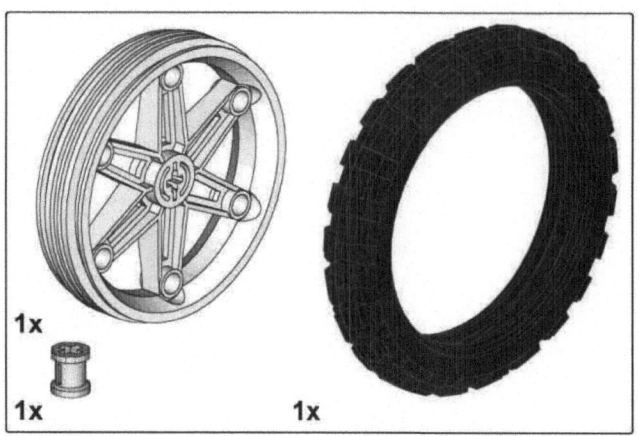

1x

1x

1x

10

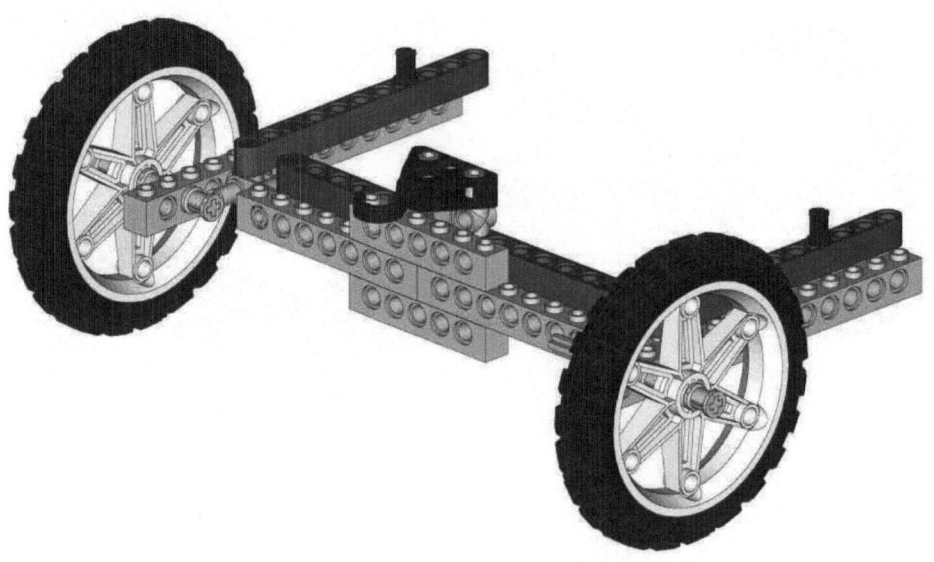

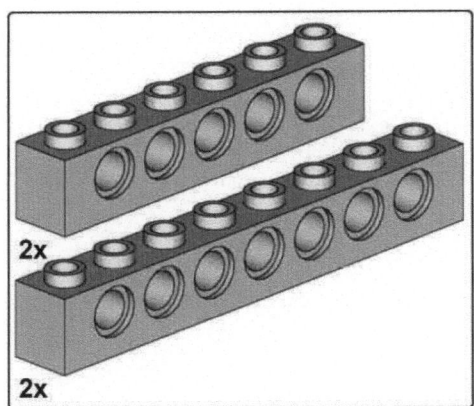

11

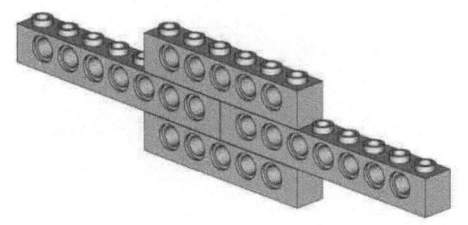

12

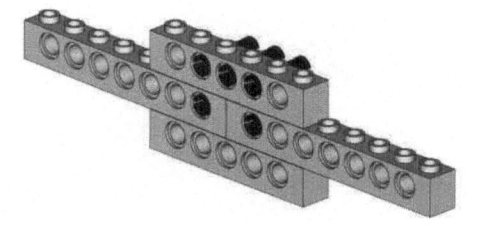

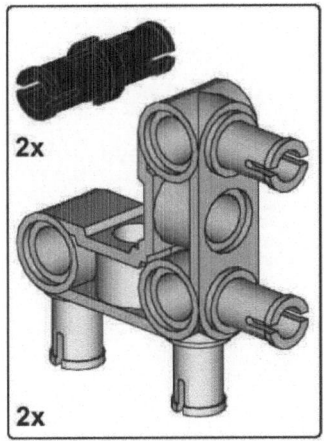

13

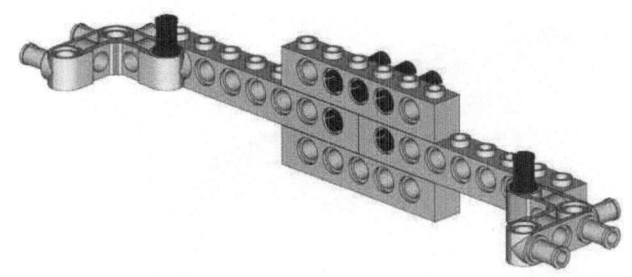

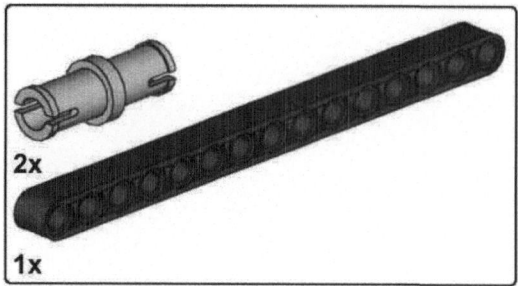

14

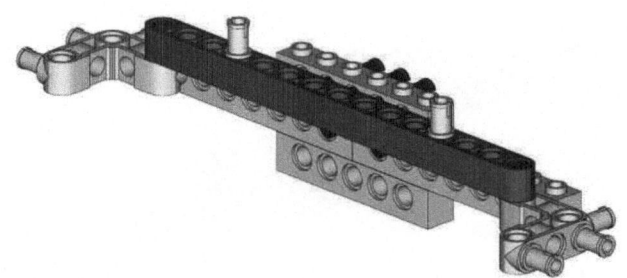

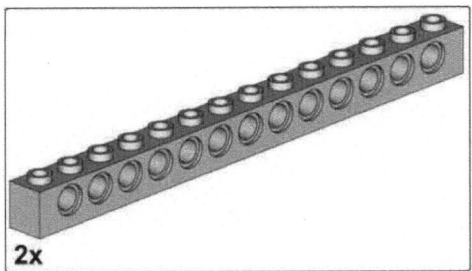

15

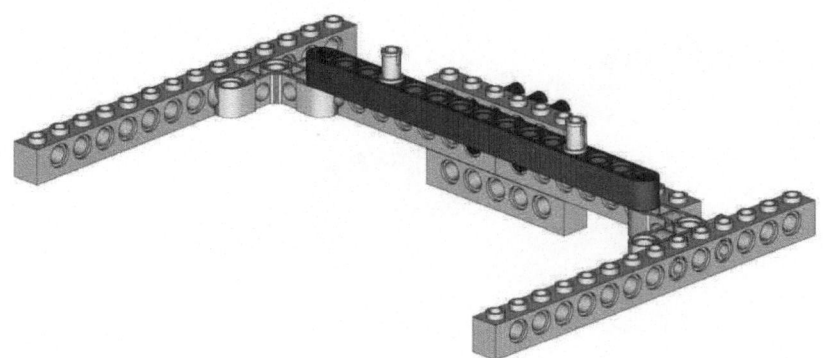

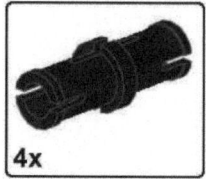

16

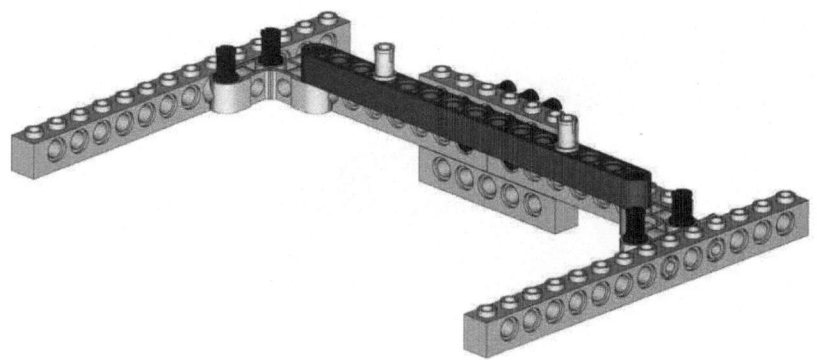

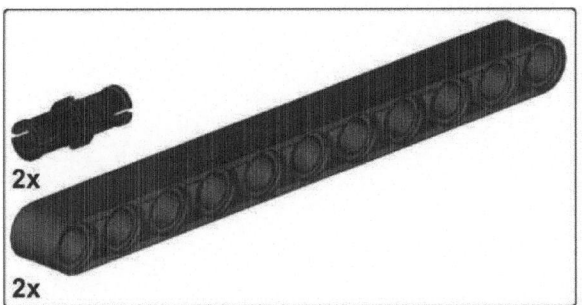

17

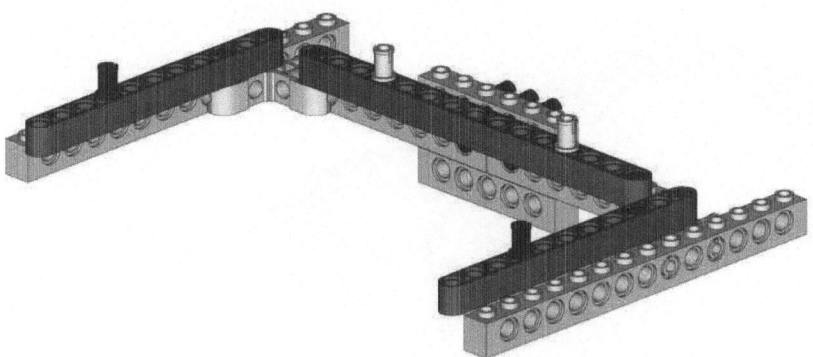

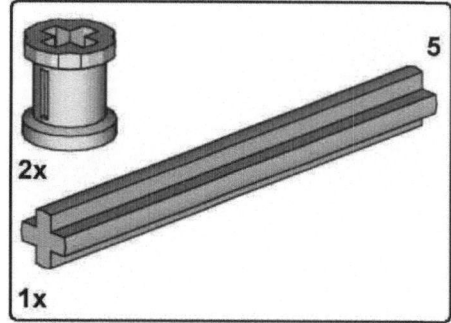

18

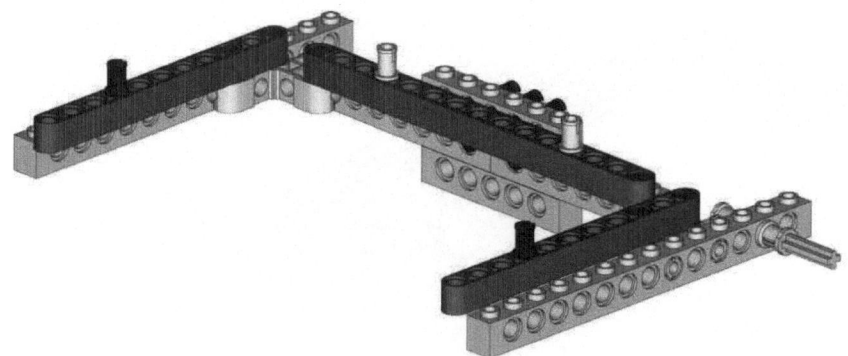

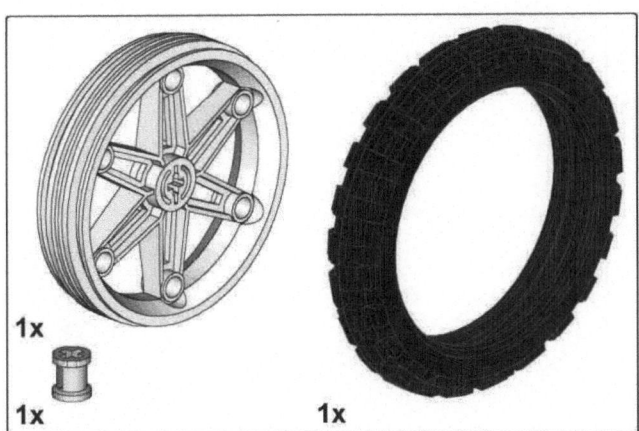

19

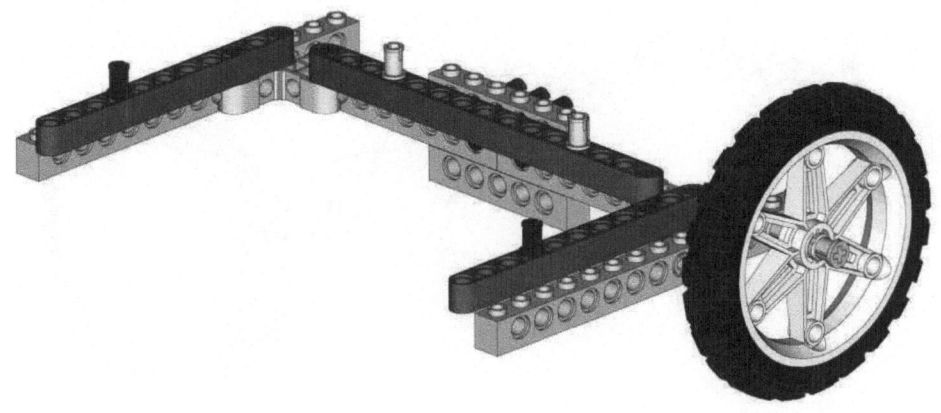

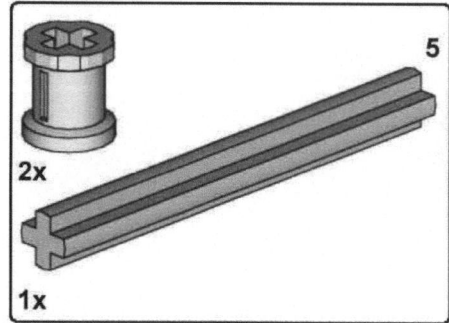

20

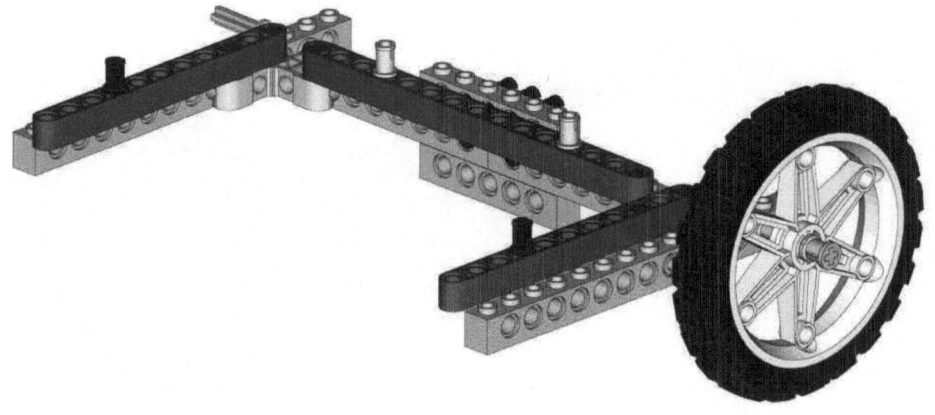

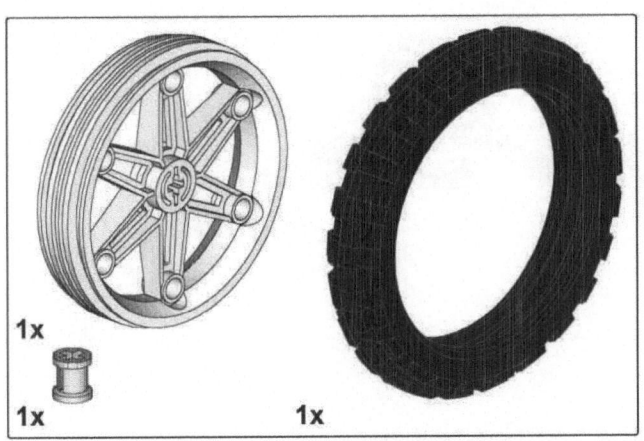

1x

1x

1x

21

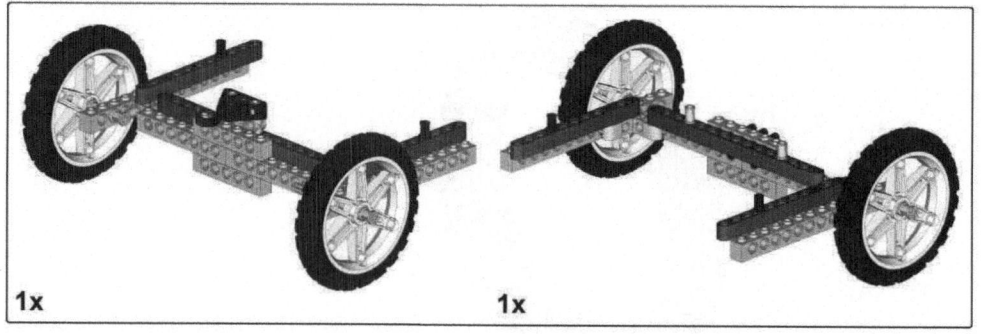

1x   1x

22

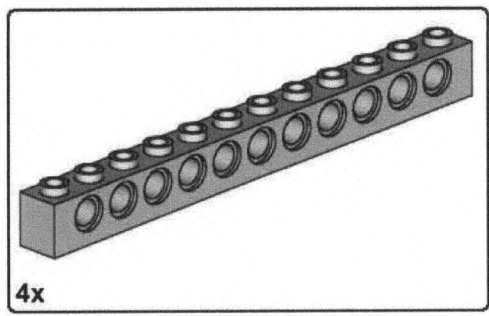

23

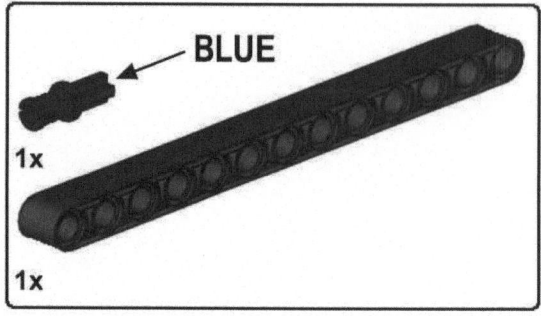

24

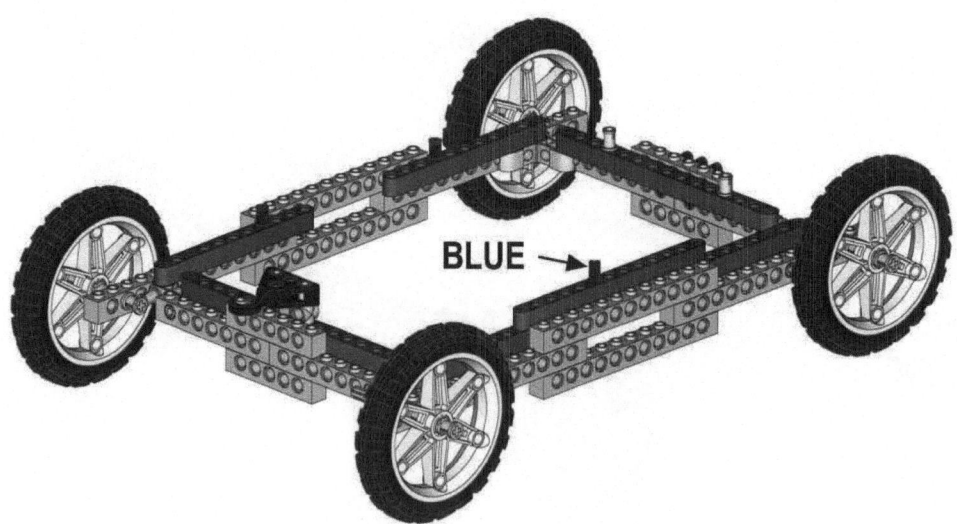

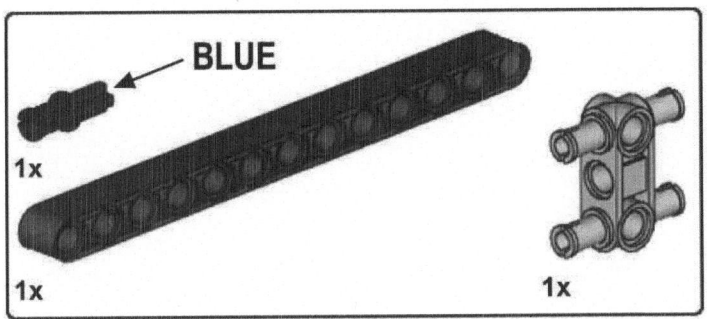

25

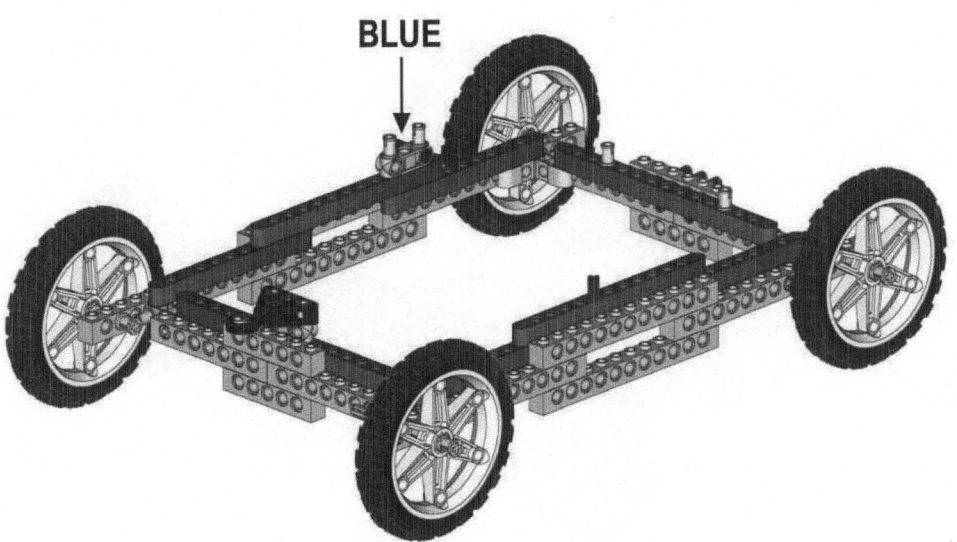

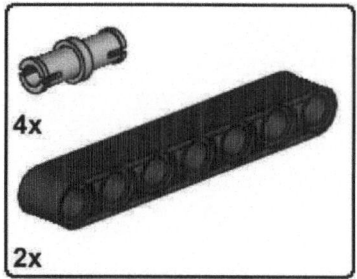

26

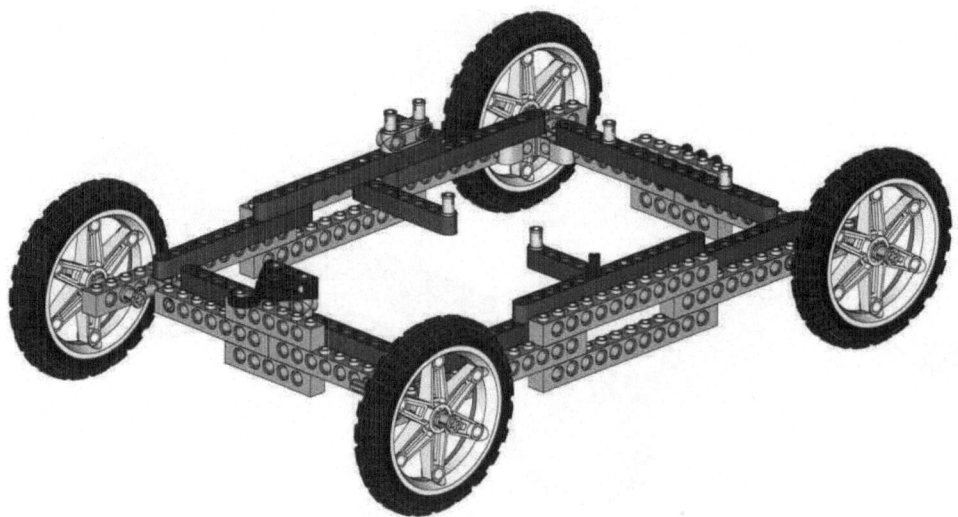

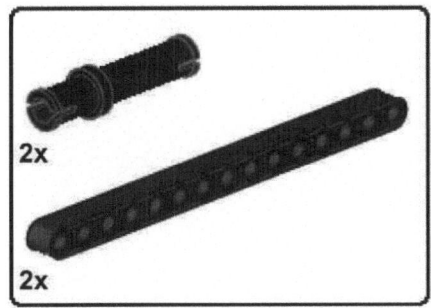

27

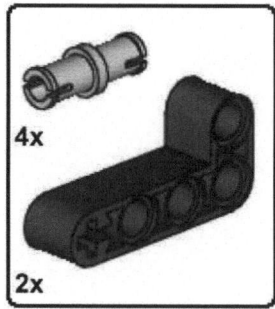

28

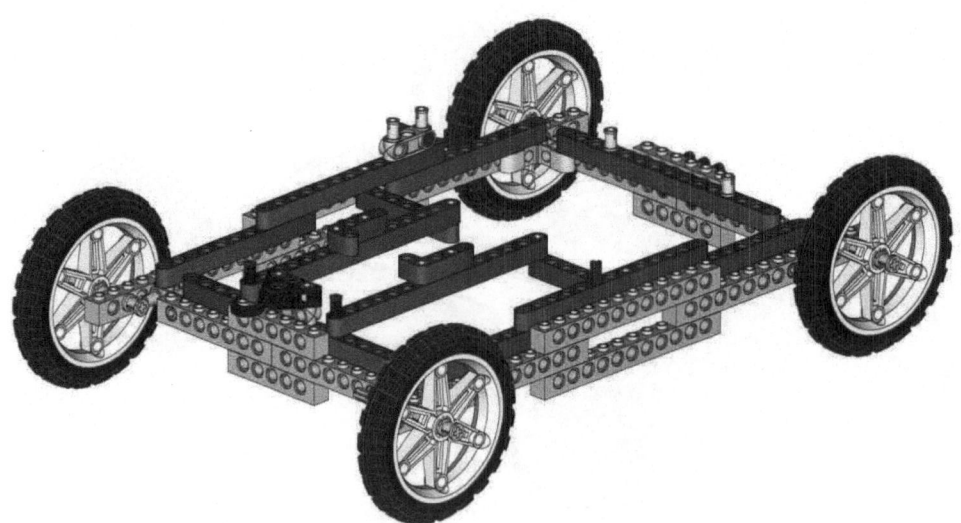

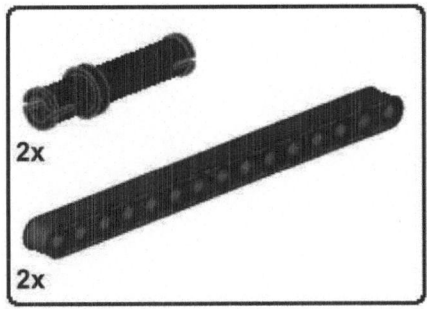

29

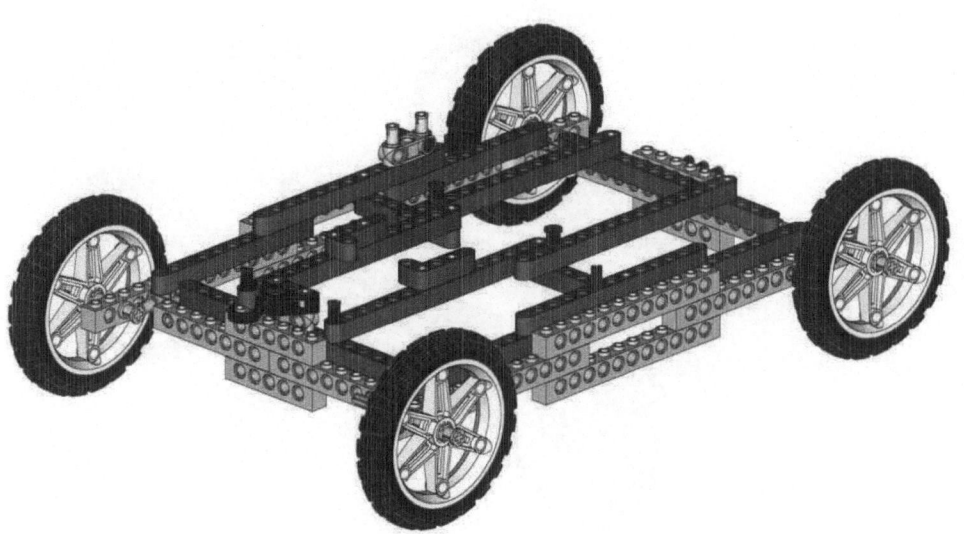

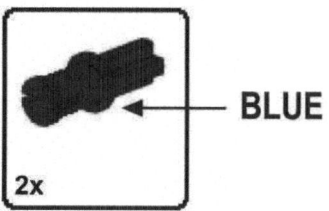

BLUE

2x

30

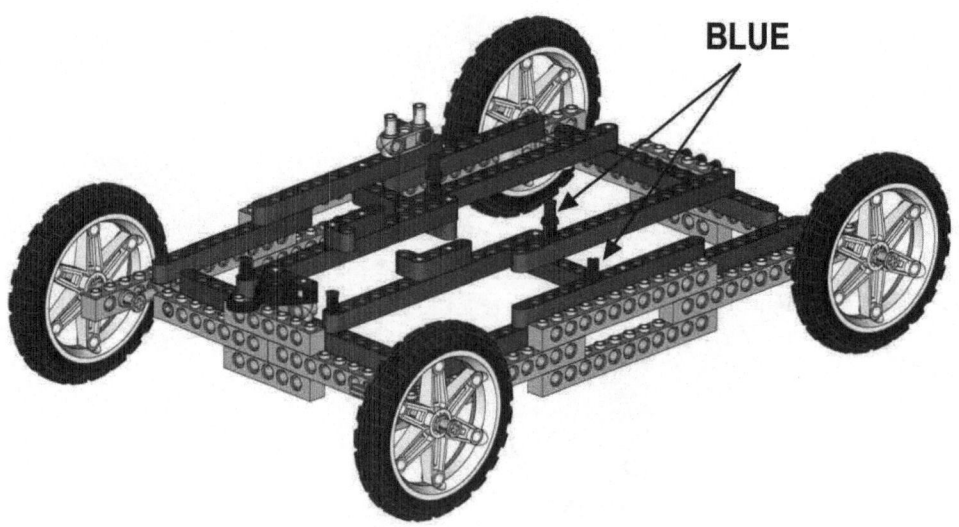

BLUE

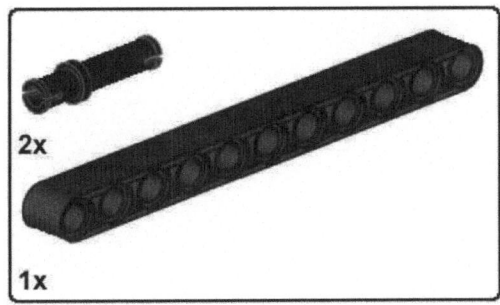

31

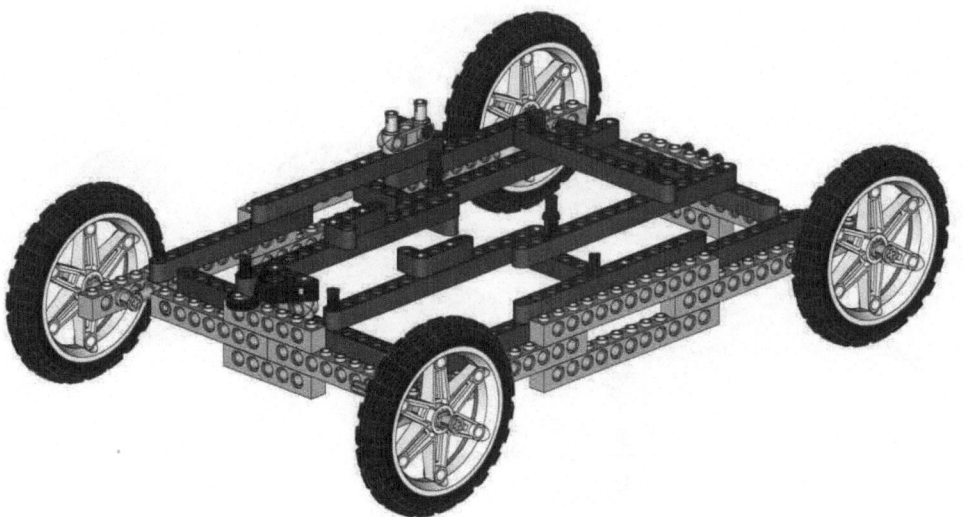

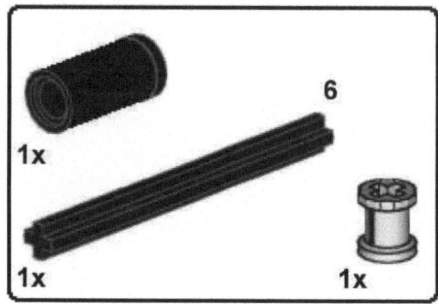

32

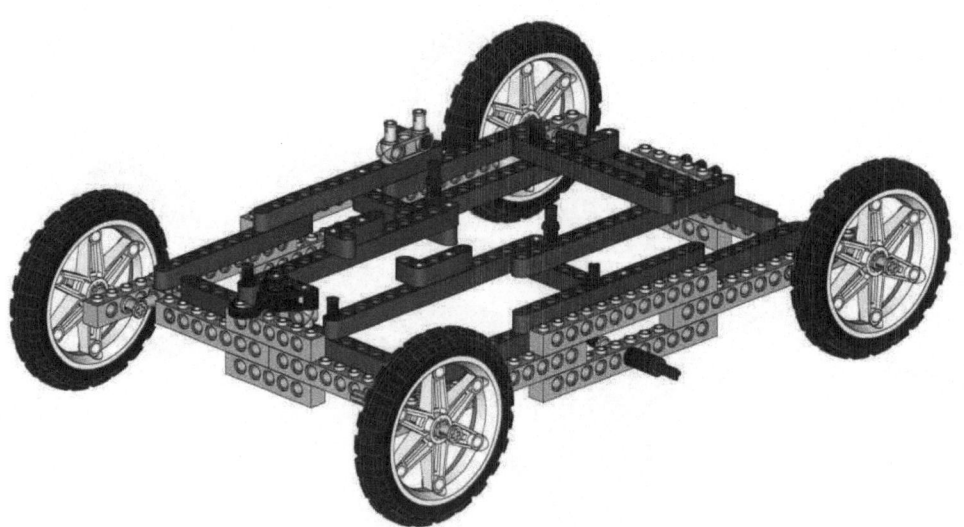

33

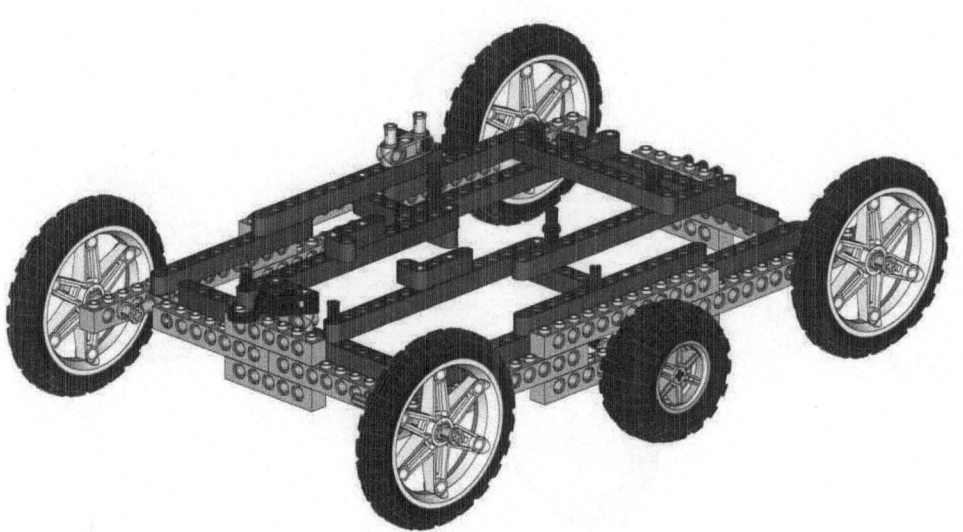

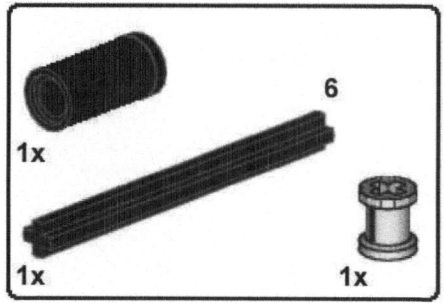

34

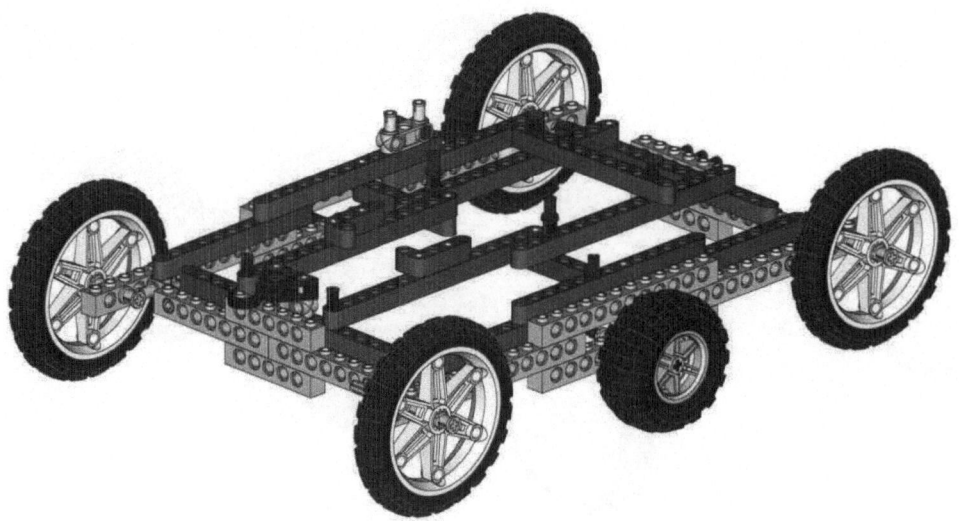

35

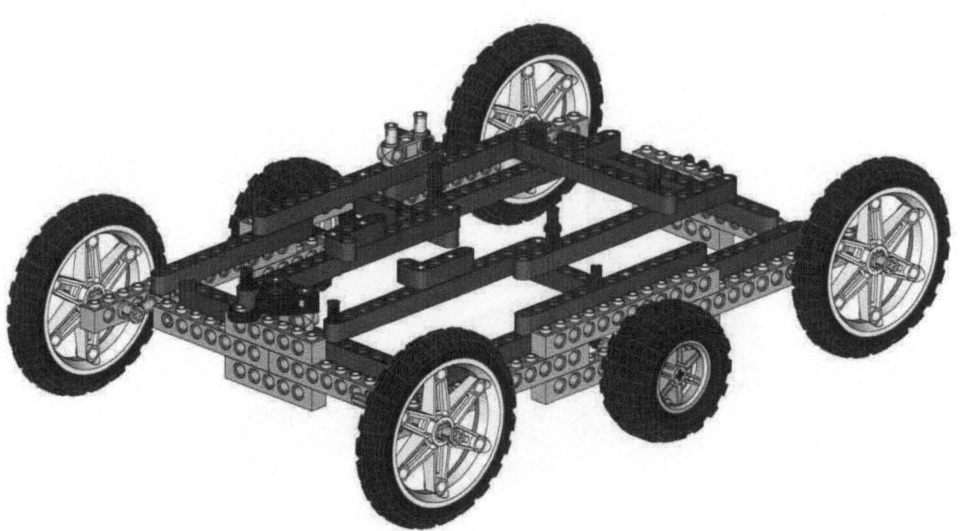

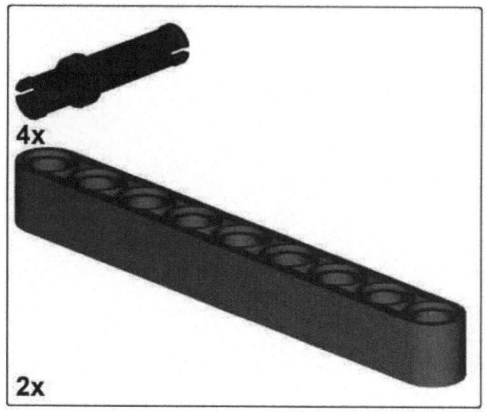

36

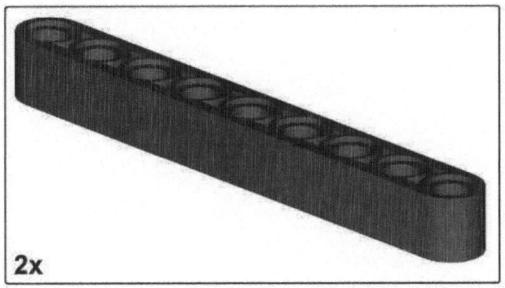

2x

37

2x

38

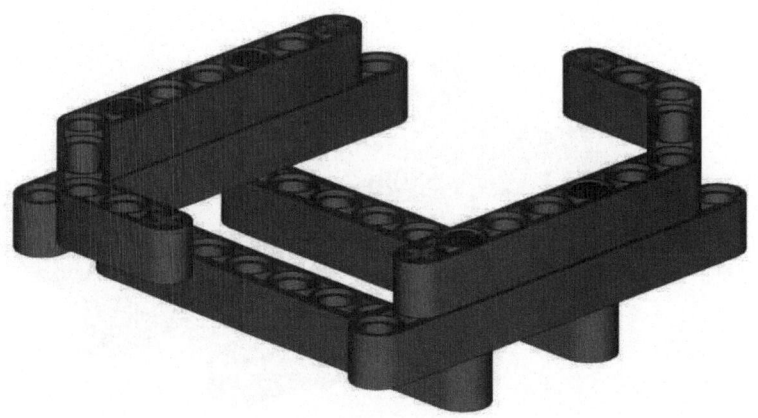

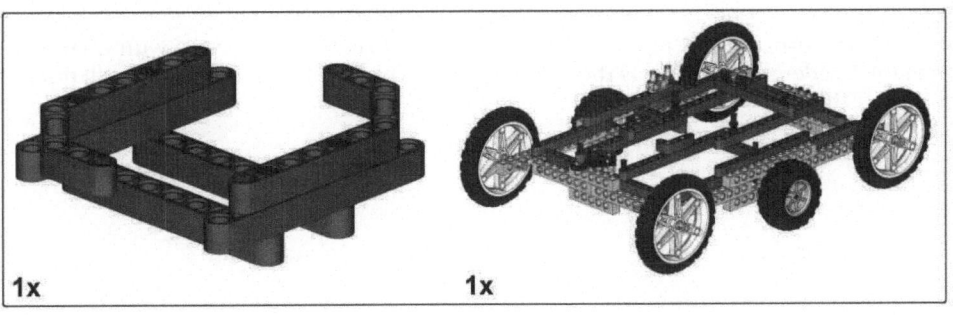

1x   1x

39

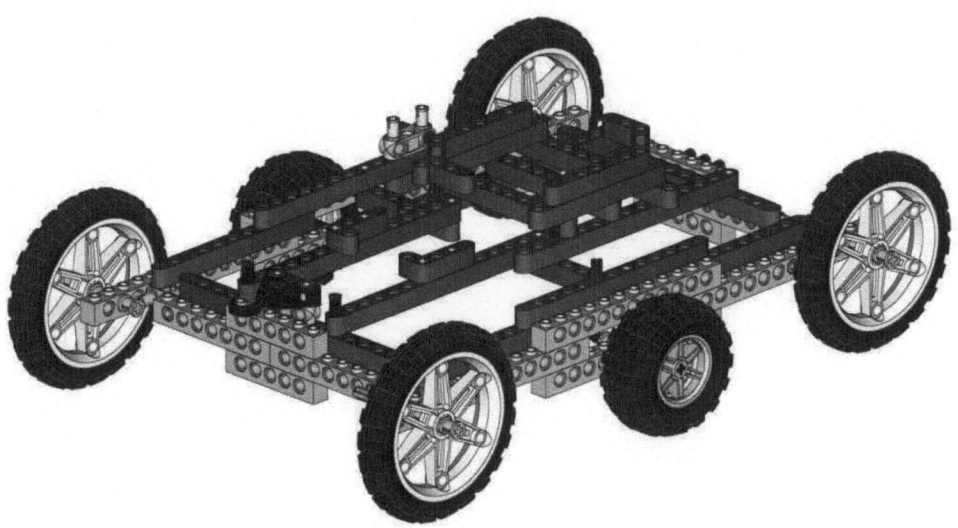

Summary

You've now got the basic form of the Lander created, so it's time to continue forward with Chapter 8 and finish up some of the Lander's tools such as the vacuum and cutting tools. When done, you'll have one Lander ready to go for third challenge.

CHAPTER 8

■ ■ ■

Finishing the Lander Mark VII

You'll find the second half of the building instructions for the lander in this chapter. Each image shows the pieces you will need to locate in the LEGO Education Resource Set and their quantities. You will also see where these pieces are to be placed.

After completing the building instructions in this chapter, you will have built the remaining subassemblies and connected them to the lander chassis. Chapter 9 will provide details on setting up the Mars Base Gamma challenge, including how to properly "damage" your lander by removing certain parts and placing them on the mat in accordance with the challenge rules.

Note   If you are uncertain about the placement of a piece in a figure, jump ahead to the next image or even go back to a previous image to make certain you've built the model correctly so far. Examine the figures carefully, and you should be able to correctly identify the pieces and their final location on the model.

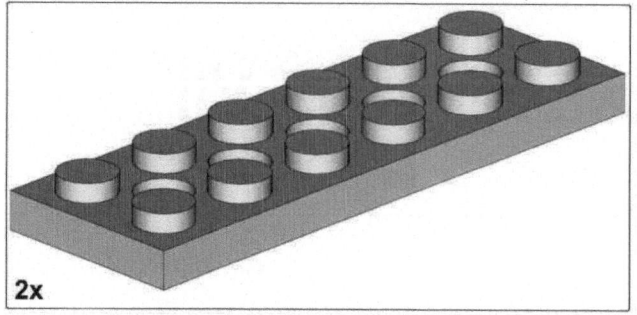

1

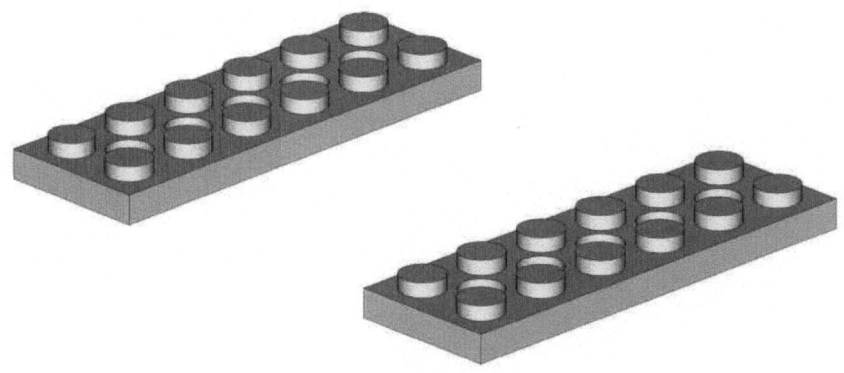

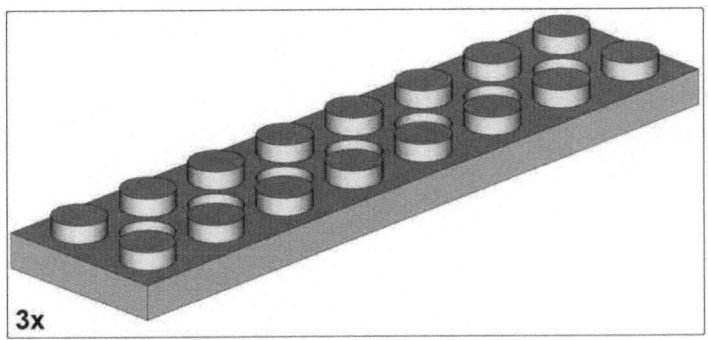

2

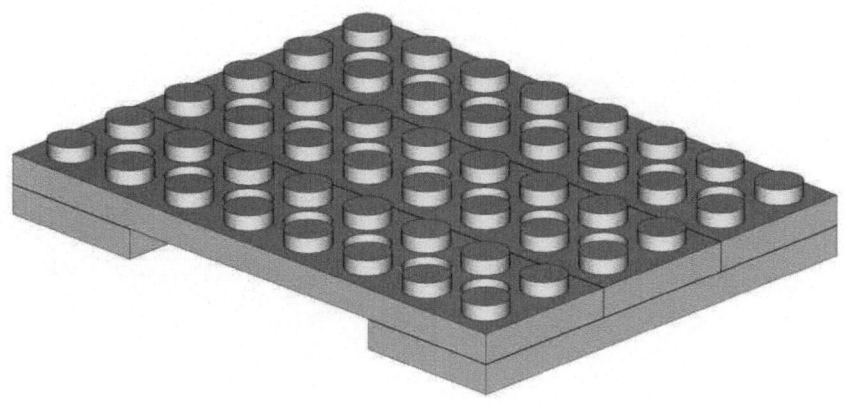

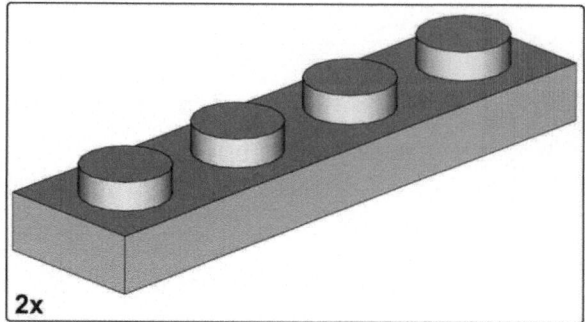

3

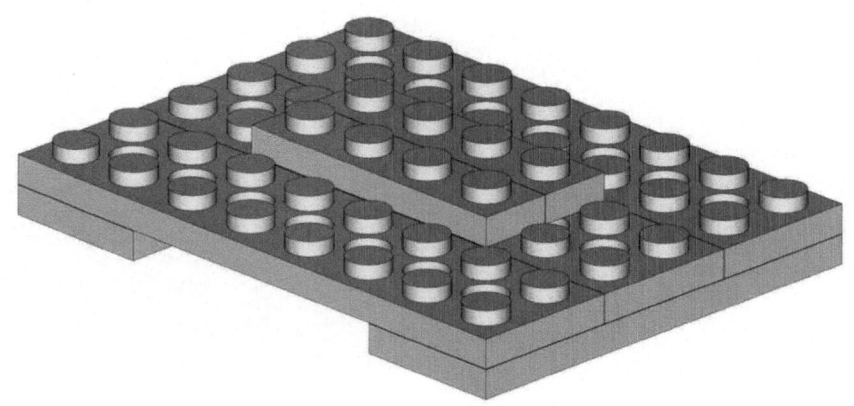

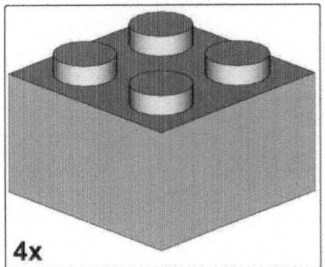

4

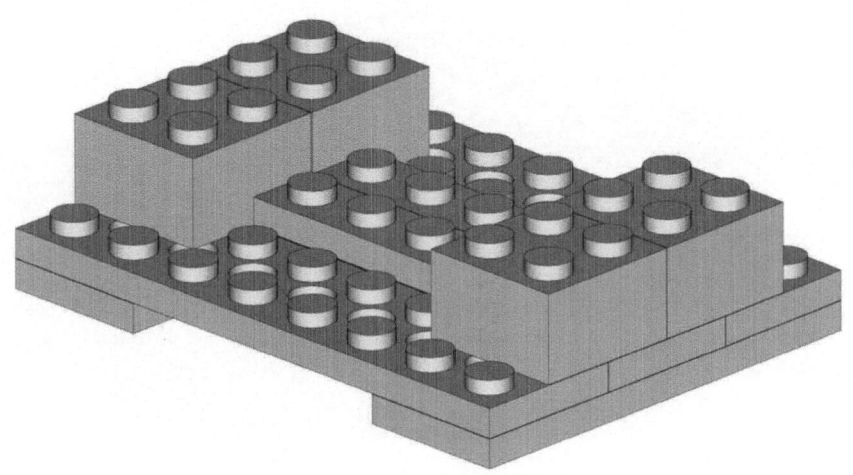

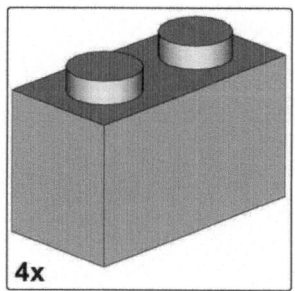

5

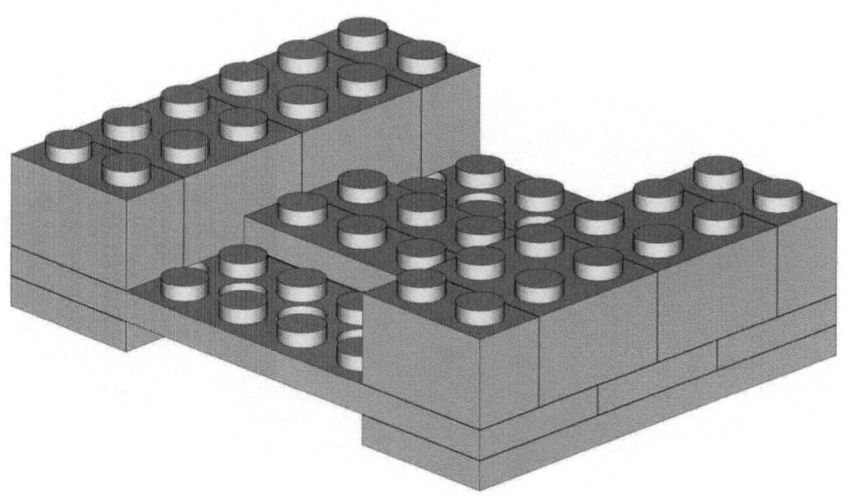

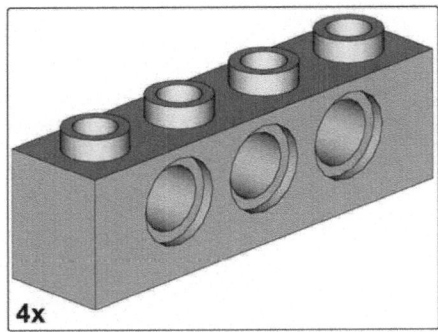

6

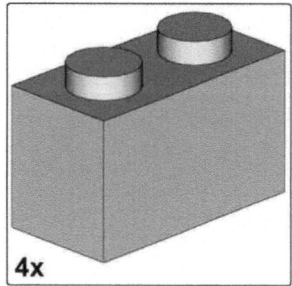

7

8

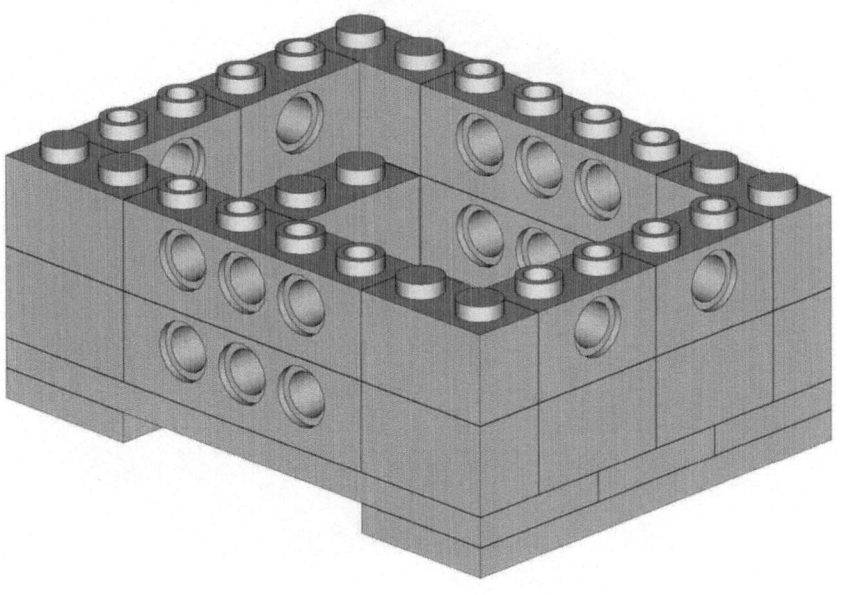

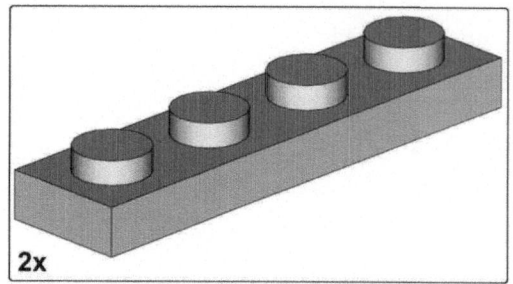

9

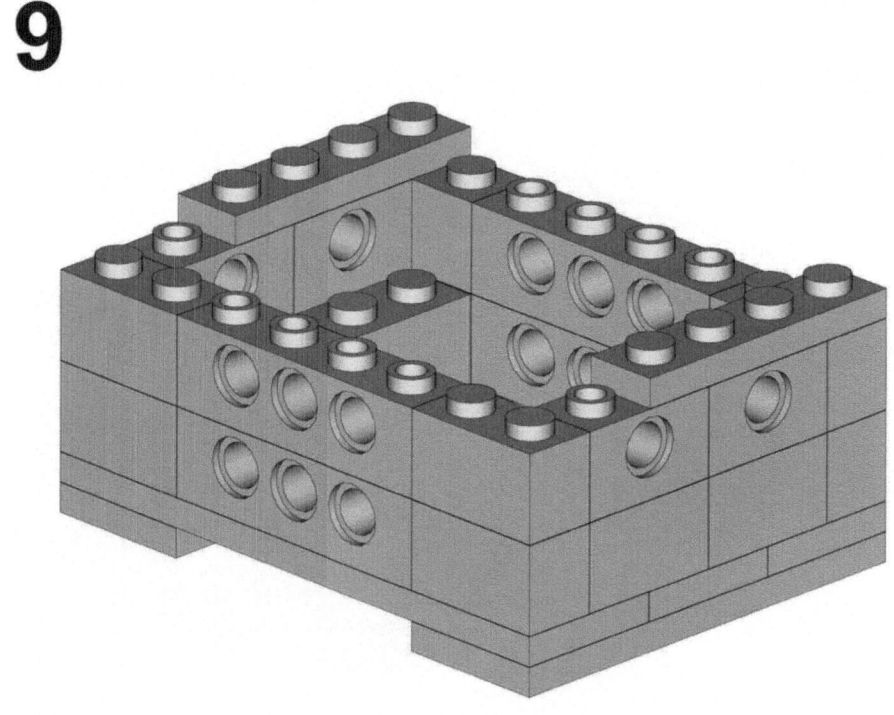

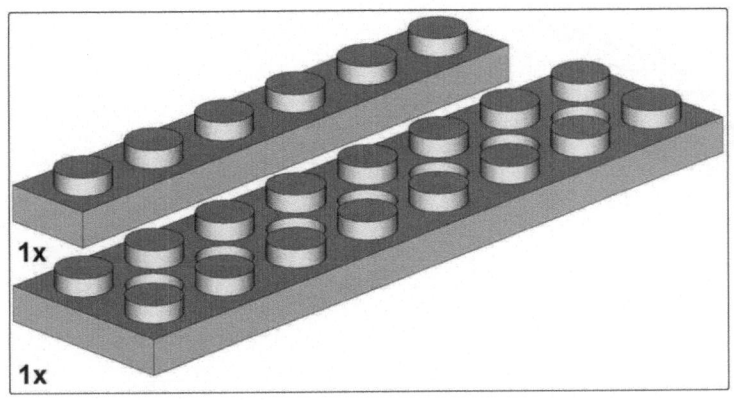

10

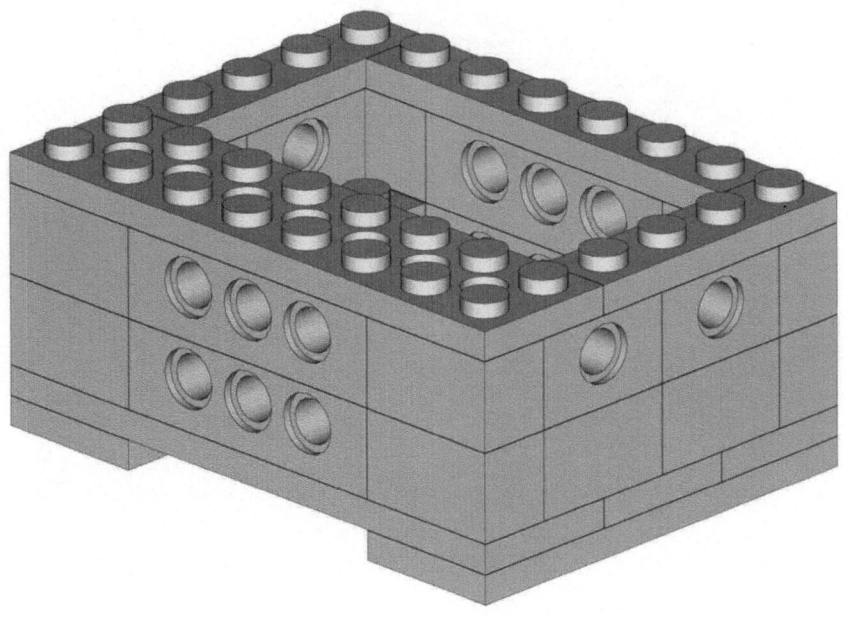

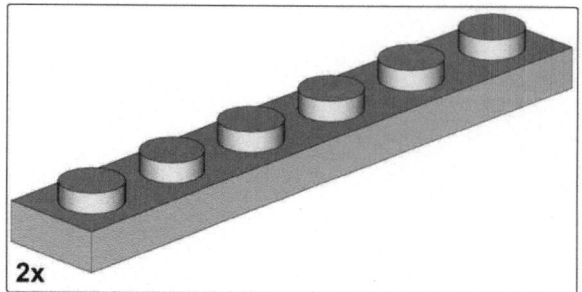

11

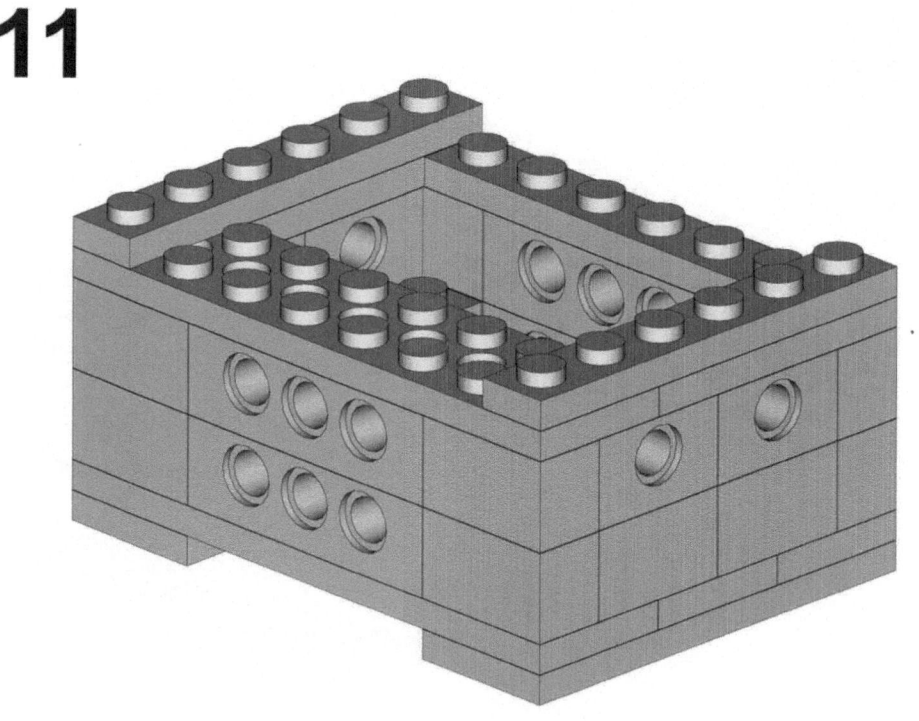

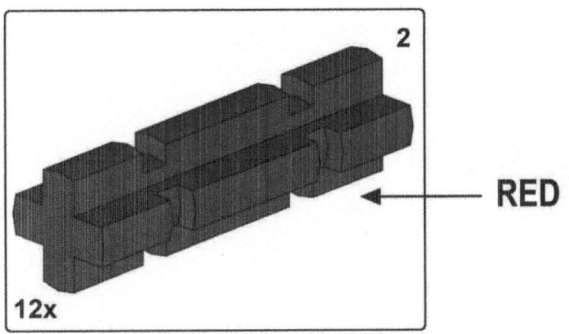

RED

12

RED

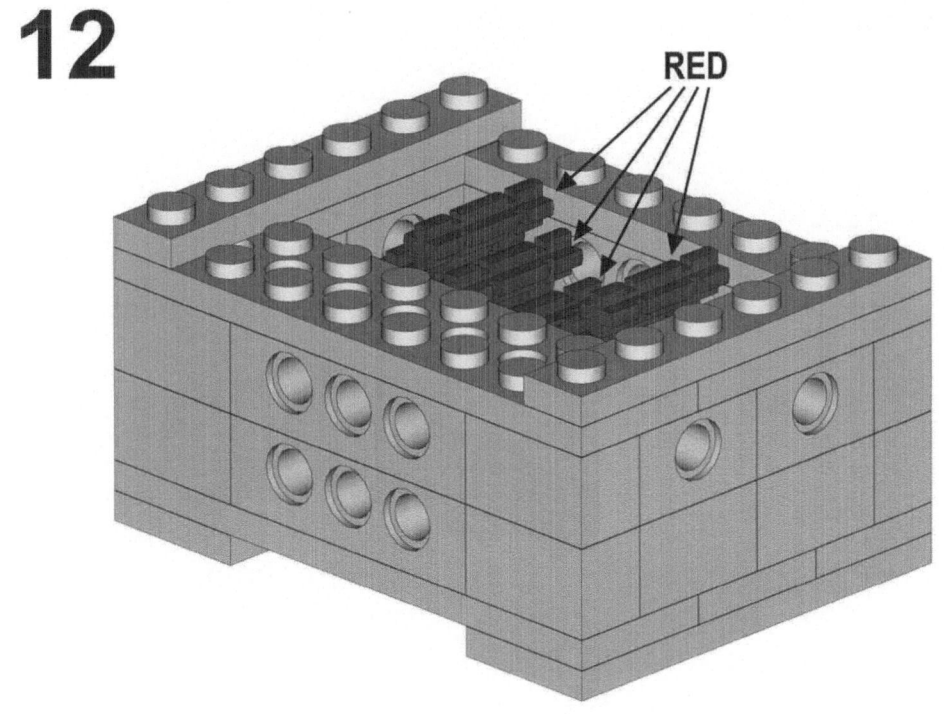

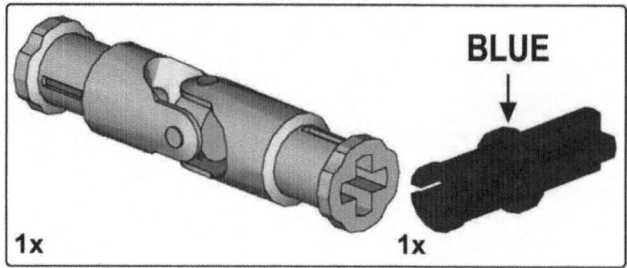

13

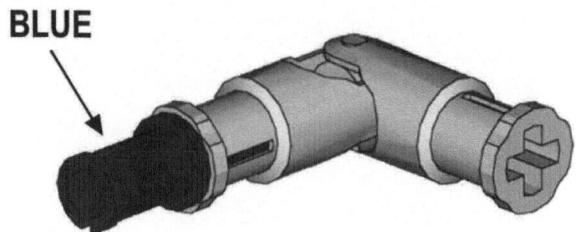

14

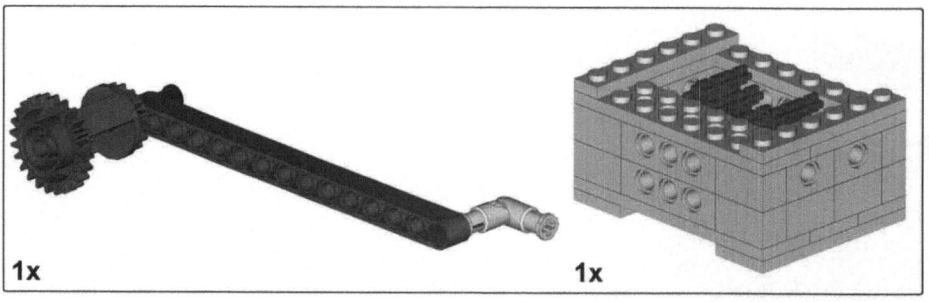

1x   1x

15

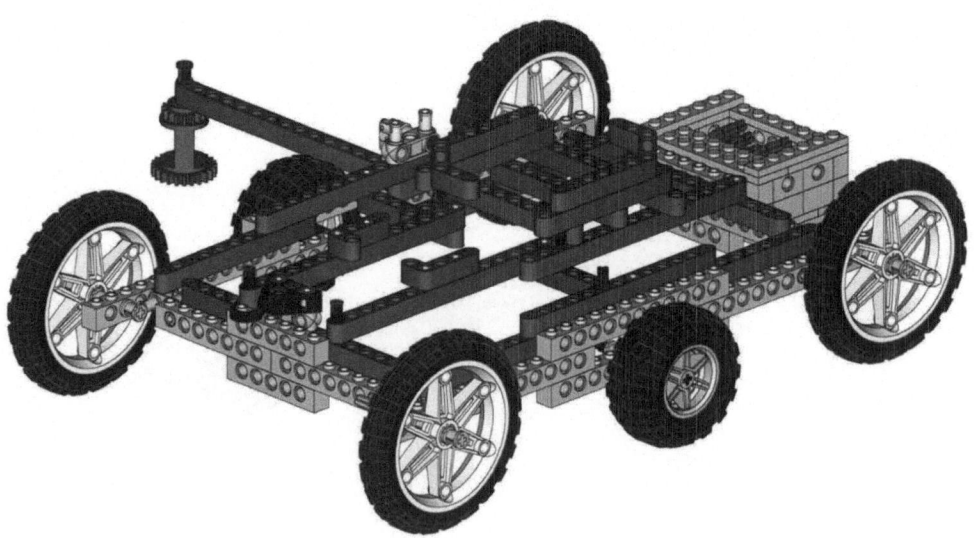

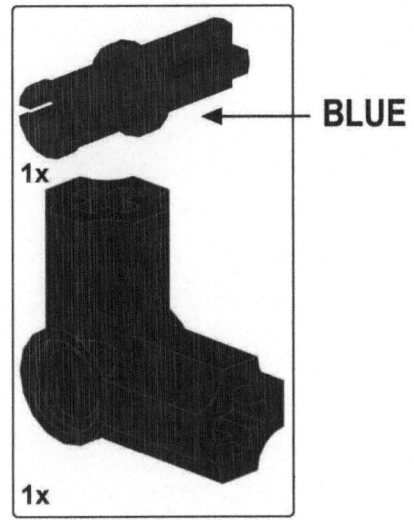

16

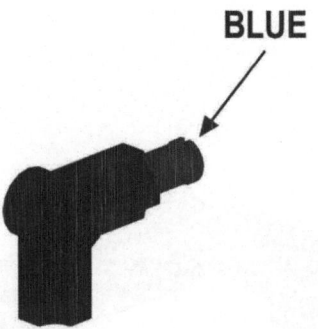

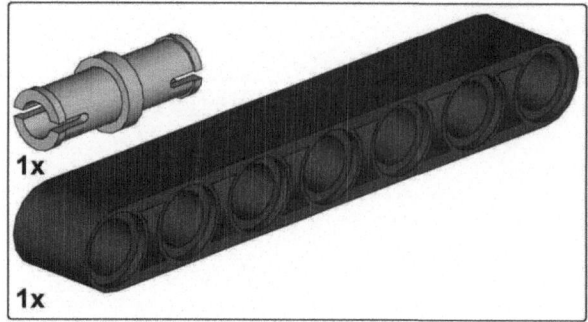

17

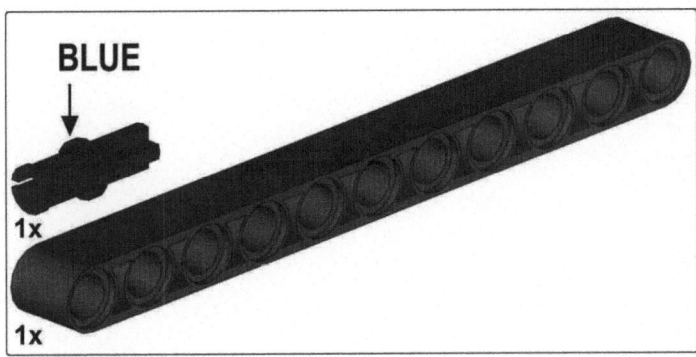

18

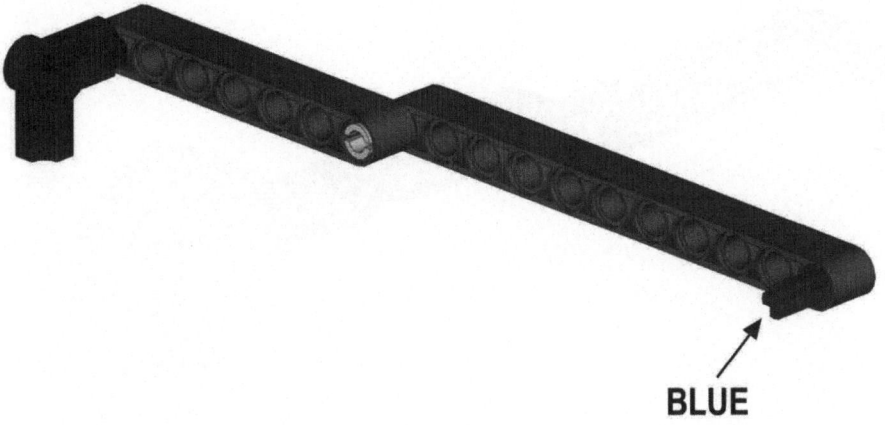

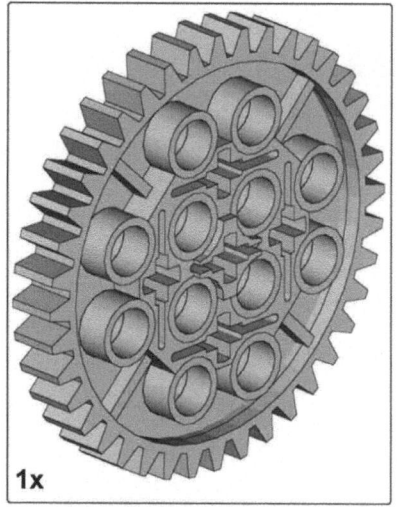

19

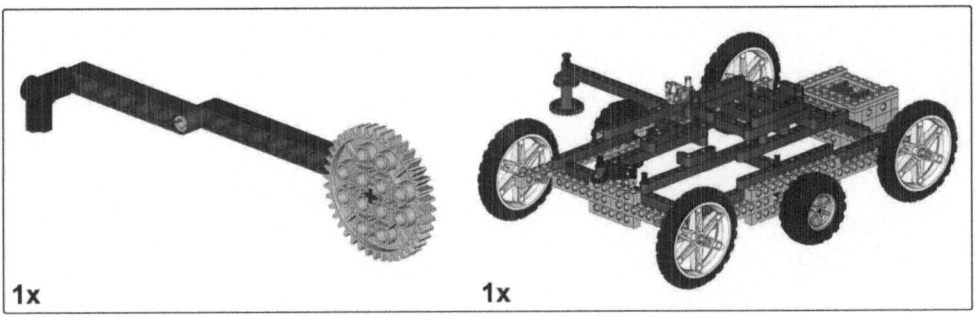

1x   **1x**

20

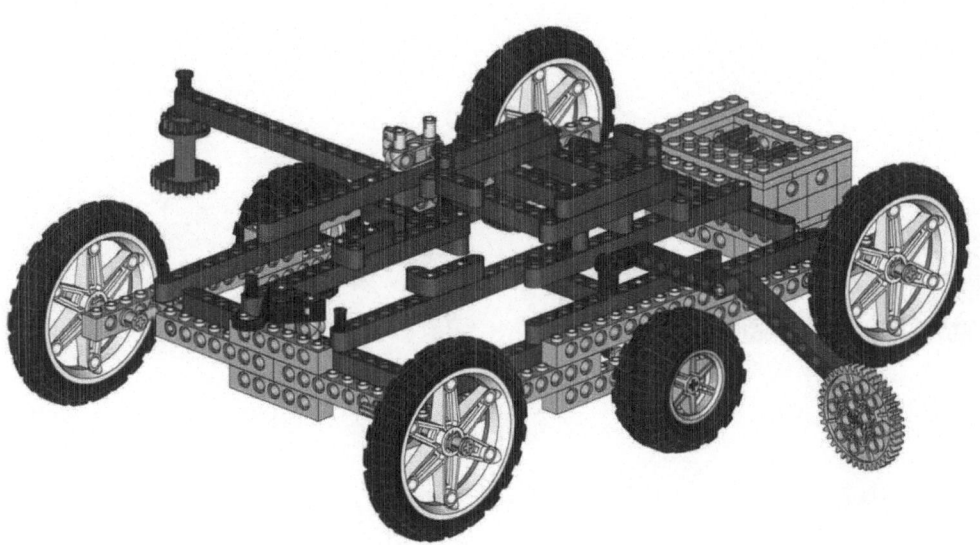

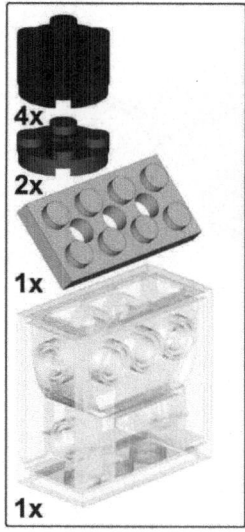

21

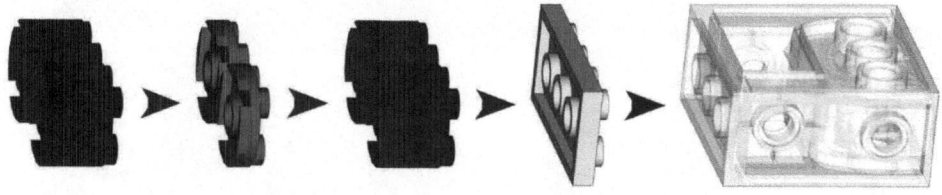

22

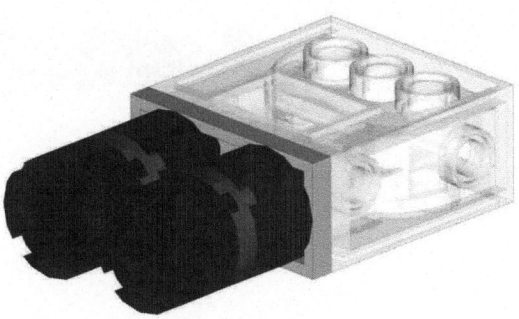

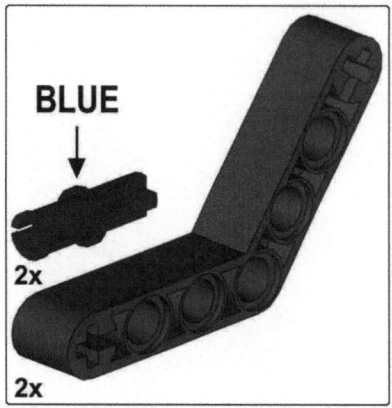

23

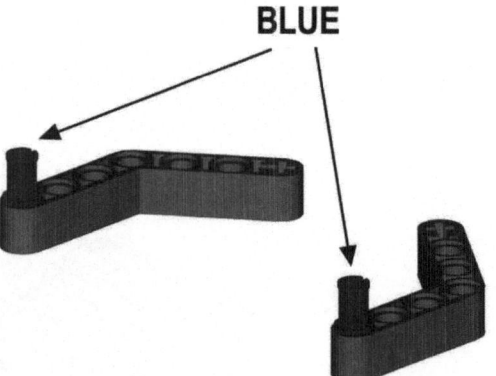

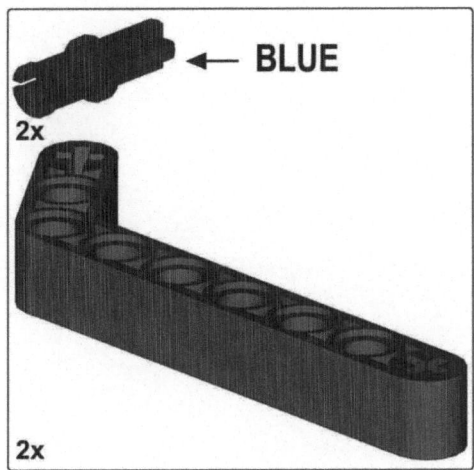

24

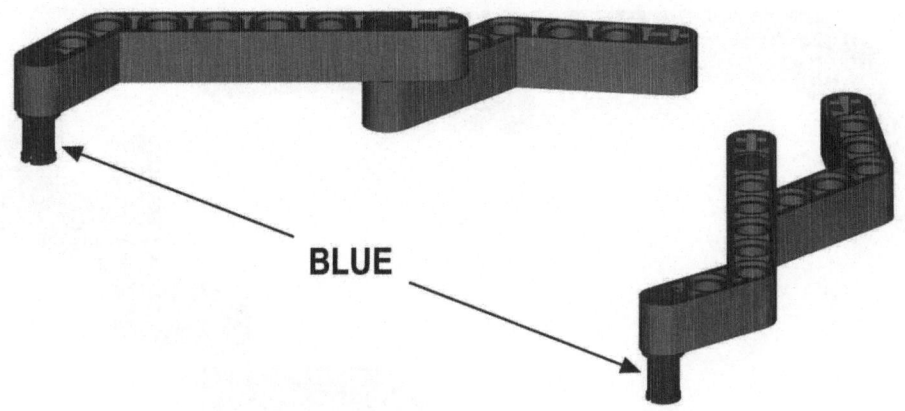

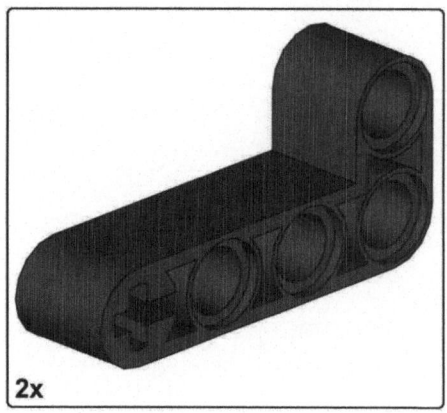

25

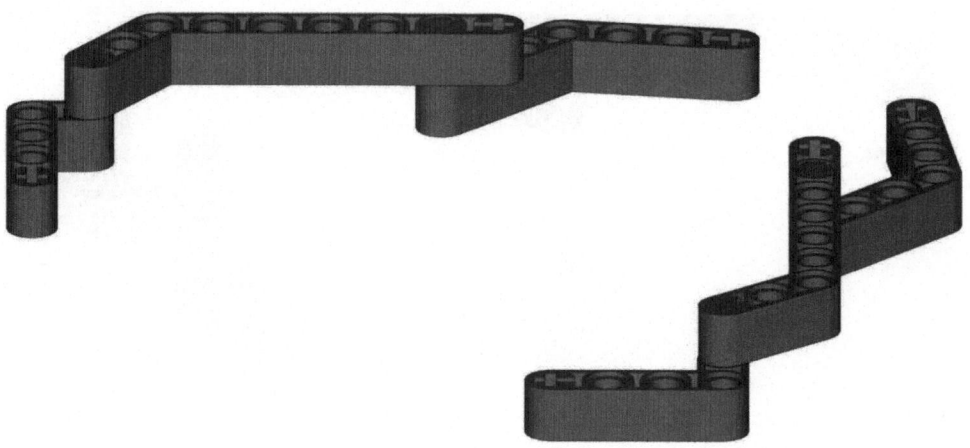

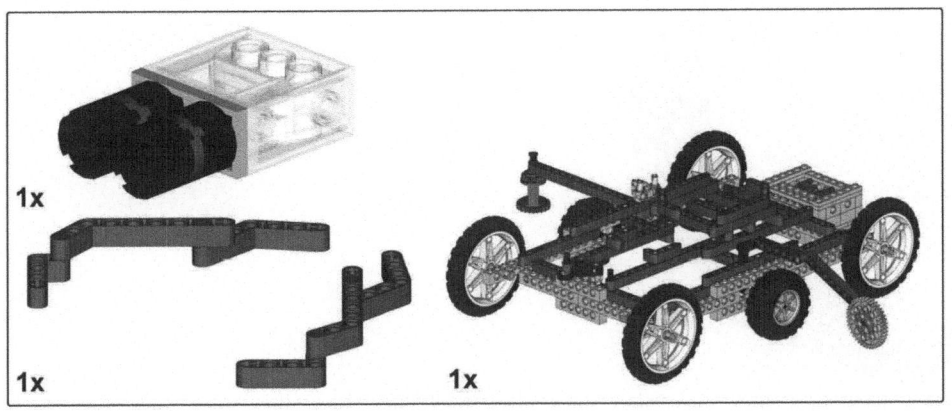

26

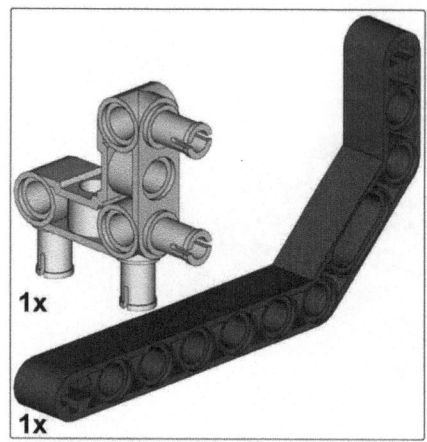

27

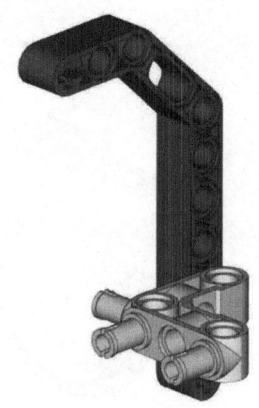

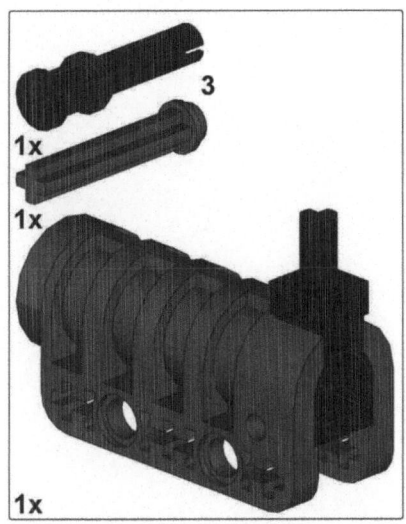

28

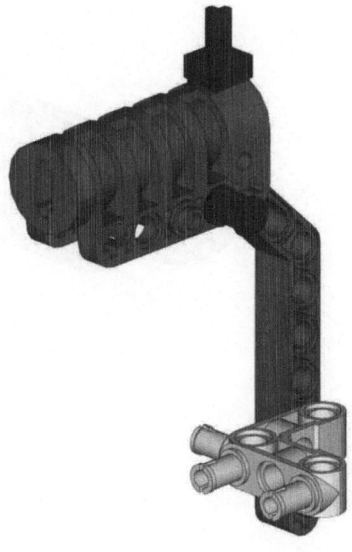

1x

29

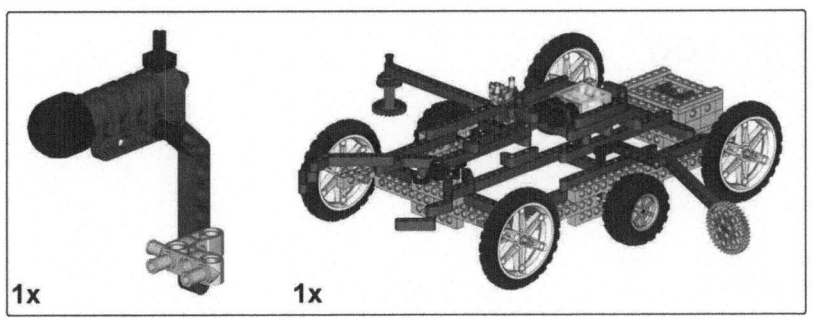

1x   **1x**

30

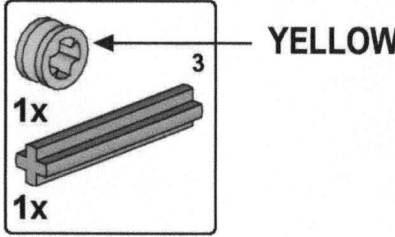

31

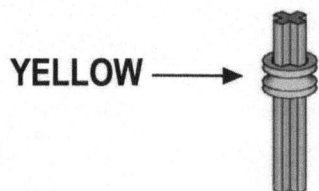

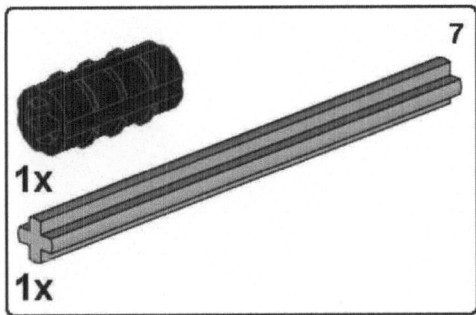

32

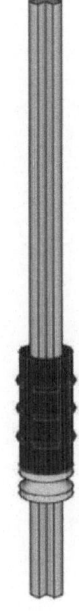

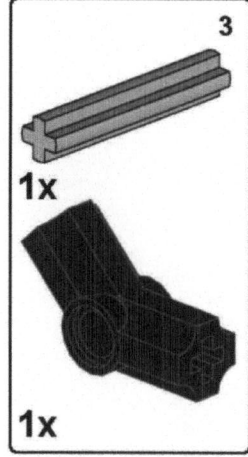

33

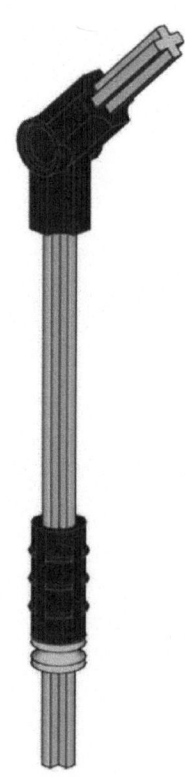

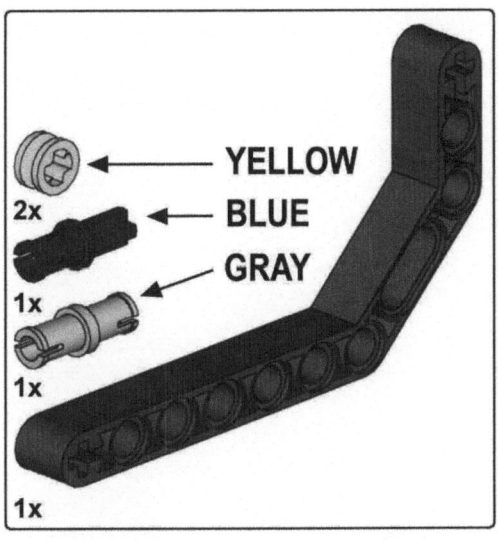

34

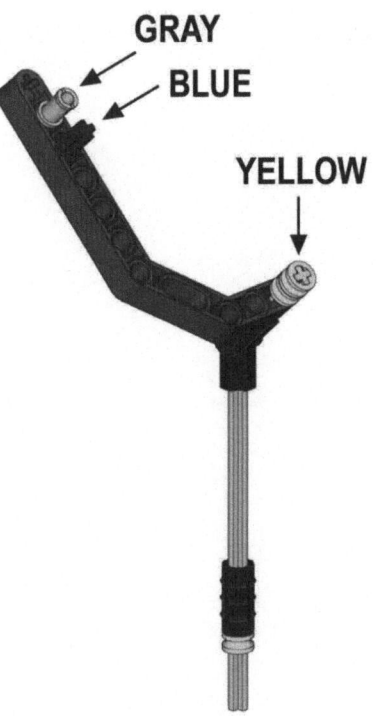

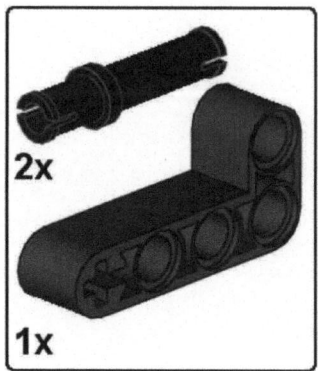

35

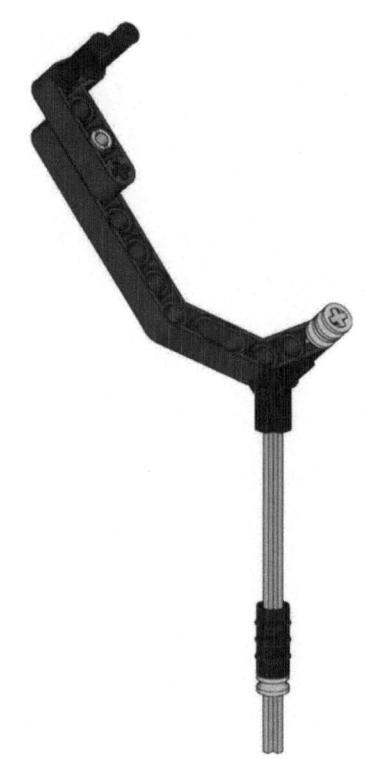

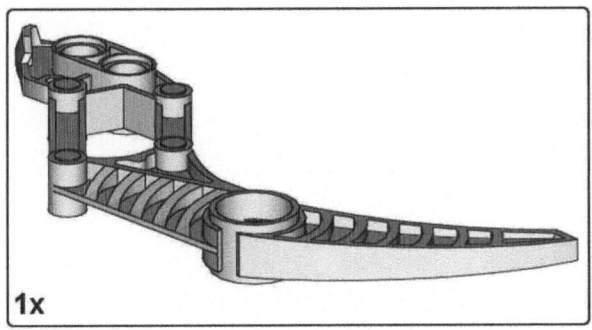

36

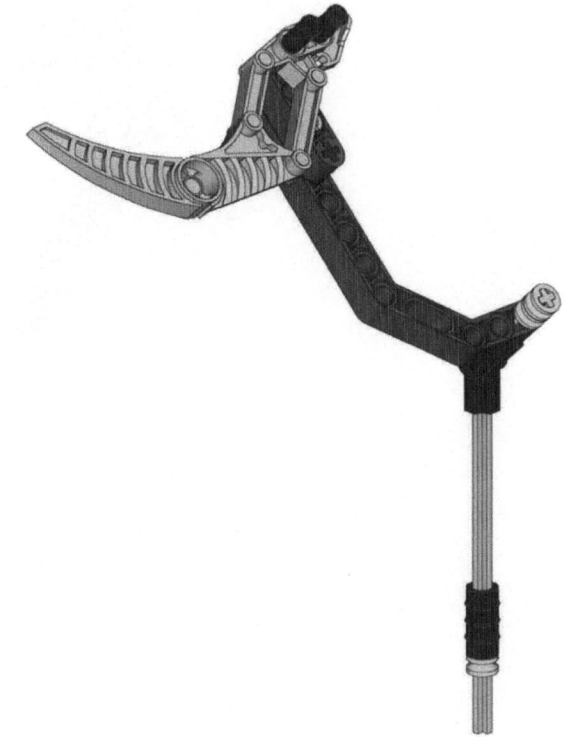

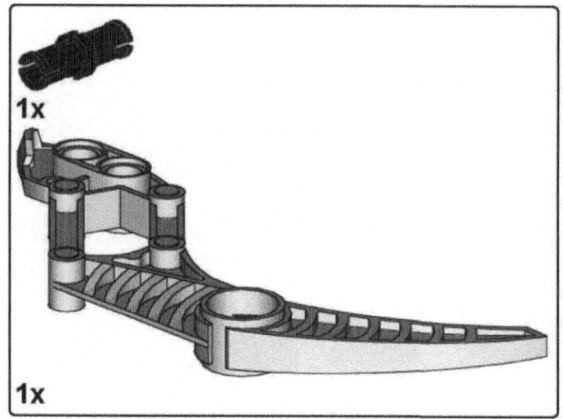

37

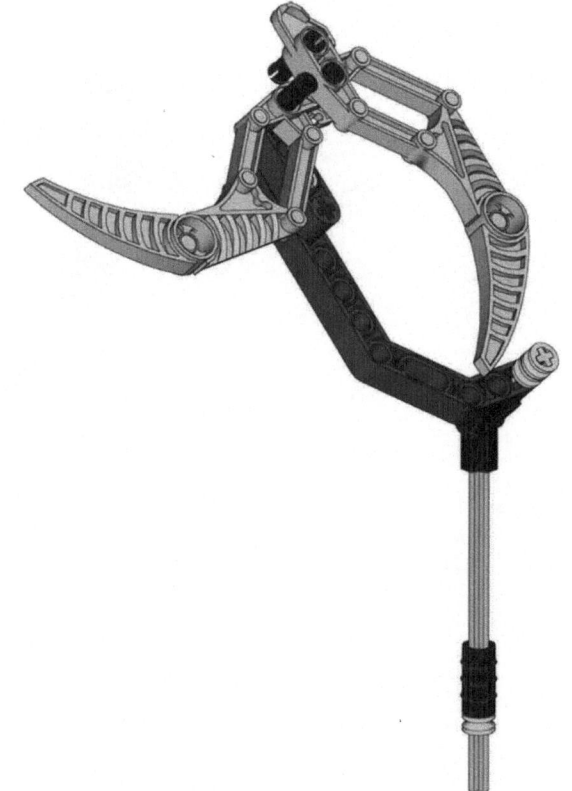

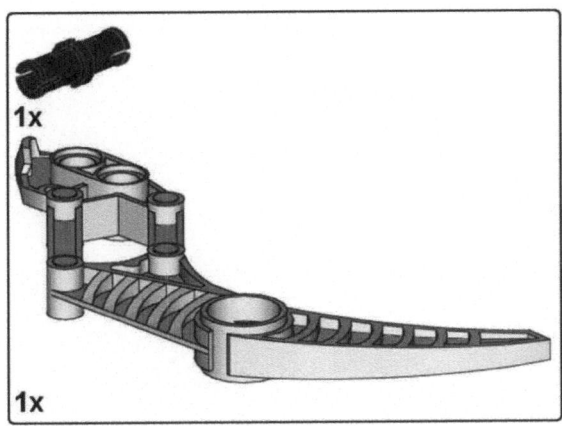

38

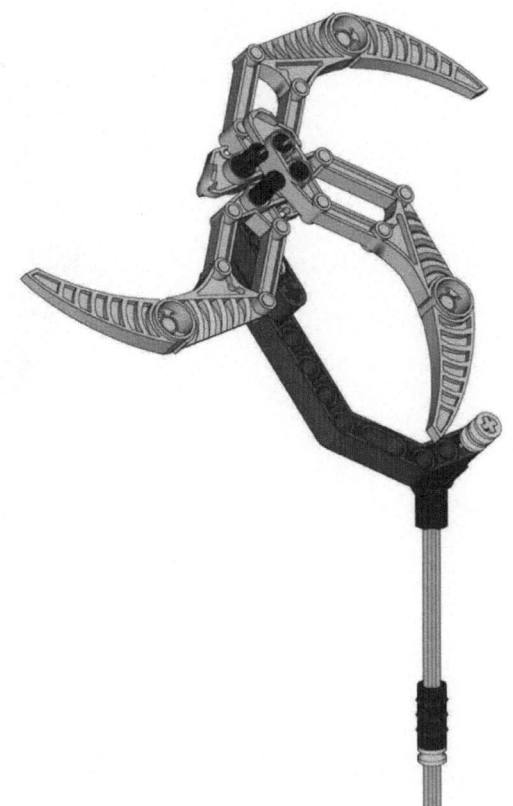

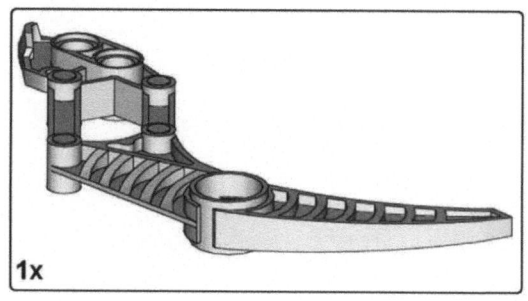

1x

39

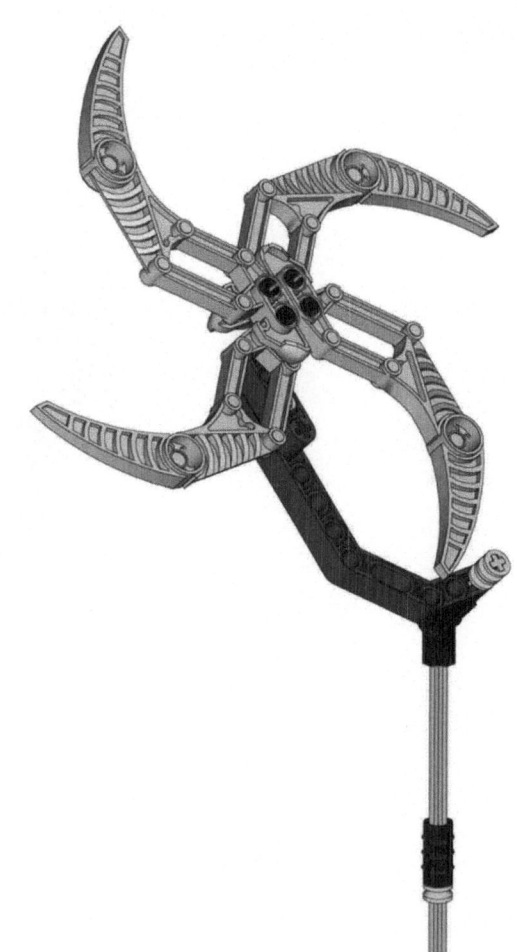

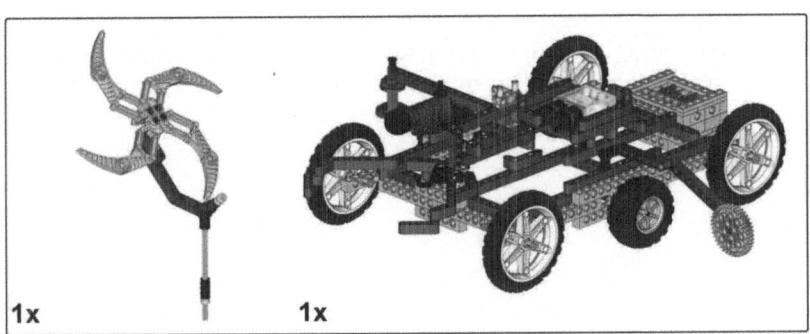

1x   1x

40

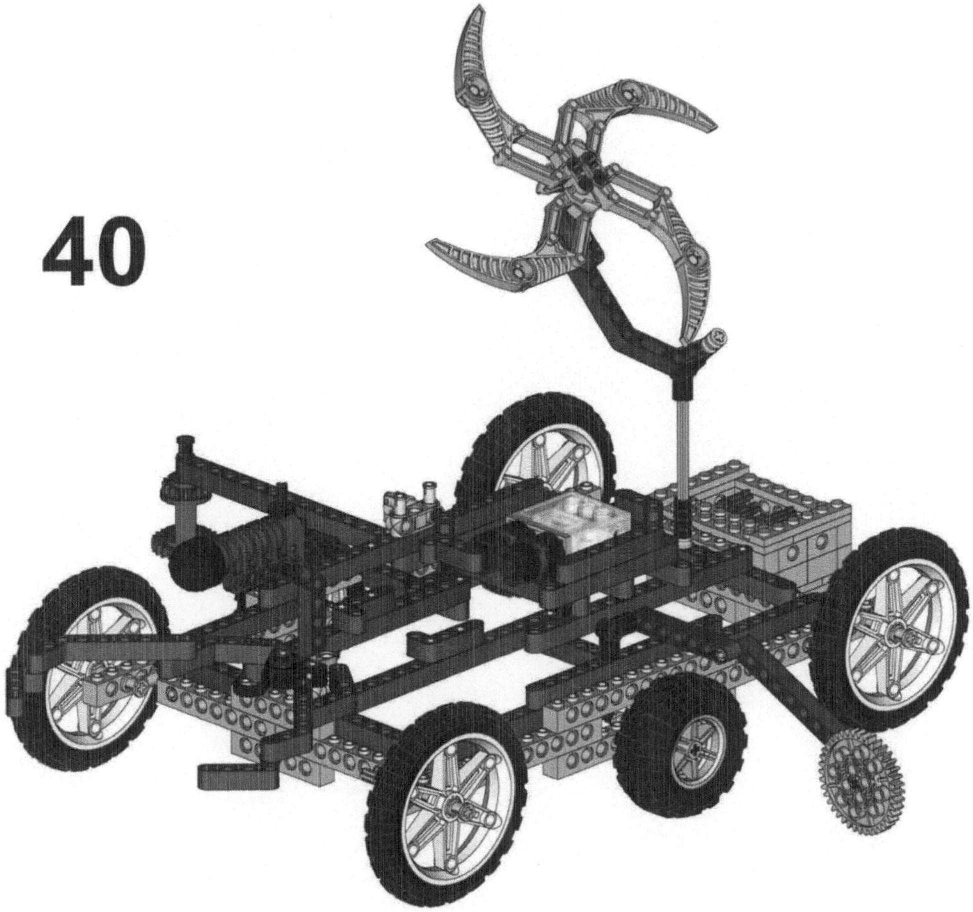

Summary

Congratulations! You've completed building the Mars Lander. But now you've got to break it down a bit to simulate the damage that it has received from the storm. Continue on with Chapter 9 to learn how to setup the challenge.

CHAPTER 9

■ ■ ■

Synopsis and Rules of the Storm Front Challenge

The Mars Base Gamma: Storm Front challenge will put your robot-building skills to the test as you attempt to rescue a damaged Lander Mark VII. The lander has fallen into a small crater and sustained damage that is preventing it from exiting the crater on its own power. Your job will be to build and program a rescue robot that can assist the lander in exiting the crater and retrieve as many of the damaged components as possible. An additional component to the challenge will be to try and retrieve the lander's rock samples that were lost in the storm.

Teams or individuals must build and program a LEGO MINDSTORMS NXT robot to successfully interact with the damaged lander and its damaged components. Retrieval of each of the damaged components and freeing the lander from the crater are both parts of the primary mission; bonus points will be awarded for the additional retrieval of rock samples but these are not required to complete the primary mission.

To successfully complete the Mars Base Gamma challenge, your robot will need to perform the following actions:

1. Push or pull the lander and free it completely from the crater along with a minimum of one damaged component.

OR

1. Retrieve and properly place the lander's battery along with one damaged component.

2. Fire the lander's grappling gun.

Additional points will be awarded for each additional damaged component retrieved as well as for the rock samples.

Setting Up the Challenge Area

The Challenge Area (CA) that defines the boundaries for your robot must adhere to the restrictions described in this section. Begin by using Figure 9-1 as a guide for proper placement of the models, or download and print the Challenge Area Mat PDF file.

Note   You may download the full-color PDF and black and white PDF mat files for use in this challenge by visiting www.marsbasecommand.com and clicking the Mars Base Gamma Mat button. The files may be taken to a printer and printed as a mat (color or black and white) to be used for the challenge area. Permission to download and use the file for private use is granted. The file may not be sold, and printed mats may not be sold. *The mat is not required to run the challenge.*

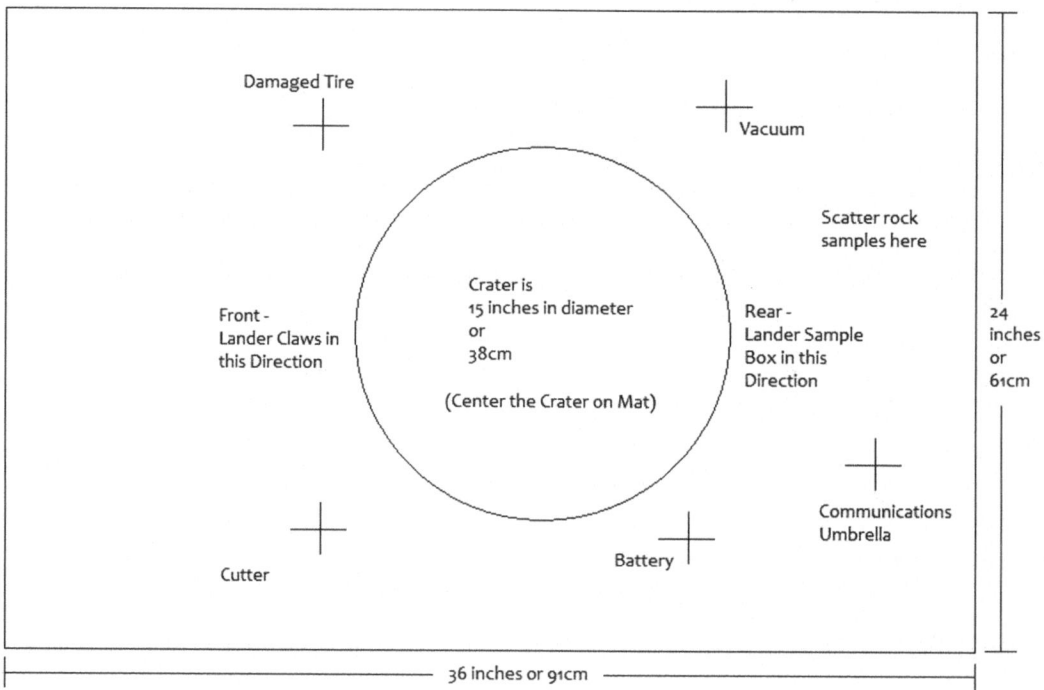

Figure 9-1. The challenge area for Mars Base Gamma

The CA must a flat surface with dimensions of 24 × 36 inches (2 × 3 feet) or approximately 61 × 91 centimeters (refer to Figure 9-1). The greater distance is the CA width.

For purposes of the challenge, the border of the CA is arbitrary; your robot may travel beyond the edges of the CA without any penalty. You may choose to enlarge the challenge area if you wish using tape or any other items.

Place the lander body directly in the center of the mat with the claws facing to the left side and the sample box facing to the right (see Figure 9-1).

Place the damaged components in the approximate locations specified in Figure 9-1. If you will be running the challenge against other teams, make a note of the exact location of each component as it is placed to ensure fairness for all teams (take a photo if necessary, or use a pencil or pen to mark where each part is placed, including its orientation).

The remaining items to be added to the challenge are the rock samples. Place the twelve red LEGO pieces in the approximate locations shown in Figure 9-1. The pieces should not be touching and should all fit within a 6-inch diameter circle.

Figure 9-2 shows a photograph of the Lander and the other components placed on the printed Challenge Area mat.

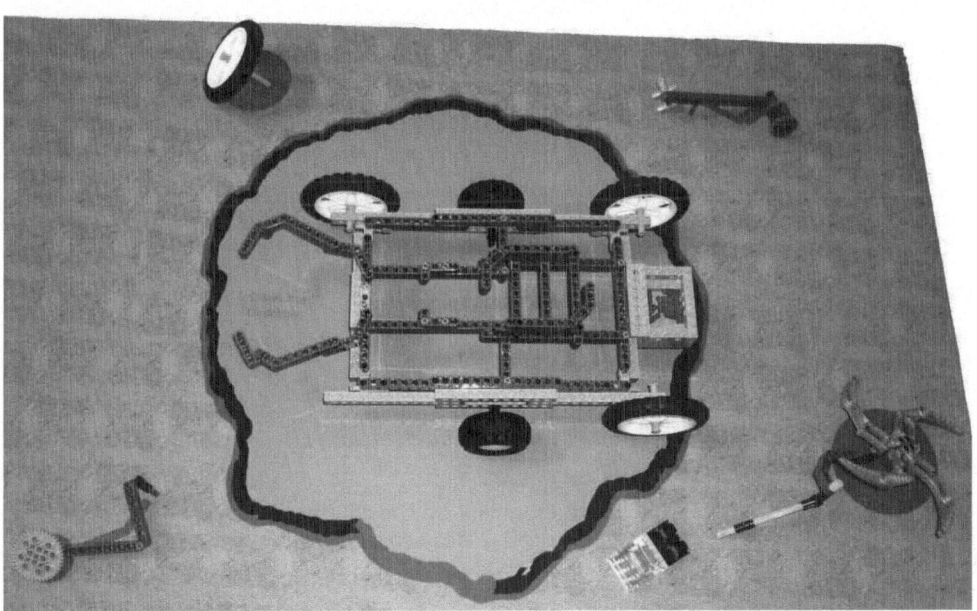

Figure 9-2. *The lander and components placed on the challenge mat*

Understanding the Challenge Rules

Several rules must be followed to run the challenge properly. If multiple teams are running the challenge, rules can be altered or removed if all players are in agreement. The rules for the Mars Base Gamma challenge follow:

1. The robot created to attempt the missions must be built using parts from a LEGO MINDSTORMS NXT robotics kit (Retail or Education versions 1.0 or higher).

2. A minimum of one robotics kit must be used to build the robot; special commendations are available for successful completion of missions using a single robotics kit.

3. A maximum of two robotics kit may be used to build the robot.

4. If a LEGO MINDSTORMS NXT Education Resource Set is used to build the lander and its components, any parts remaining in the Resource Set may also be used in the robot design.

5. Remaining parts from a maximum of one Resource Set may be used in the robot design.

6. Only NXT motors may be used in the robot's design.

7. Approved sensors are Ultrasonic, Sound, Touch, Color, and Light (NXT versions only).

8. The challenge time limit is 4 minutes. The challenge ends when the time limit expires.

9. Once the robot is placed outside the mat (or challenge area) it cannot be picked up until all mission objectives have been met or the time limit expires. (See the mission objectives descriptions for success/failure descriptions.)

10. Points are scored based on successfully completed mission objectives only.

11. The robot must be placed in a starting position outside the challenge area; exact location can be determined by team(s).

12. The robot may be oriented in any direction desired (from its starting position) but must be outside the challenge area (either defined by the mat or by a 2×3–foot area surrounding the lander).

13. Tape (any color) may be used and placed anywhere on the challenge area to assist the robot with navigation. It may be used to define the CA boundaries as well as to provide lines to follow or points on the CA. There is no limit to the amount of tape that may be used.

14. Tape, Velcro, or other items may not be used to secure the lander or its components to challenge area.

Understanding the Mission Objectives

The primary mission objective can be attempted before or after any of the optional missions. However, the optional mission objectives will only be scored if the lander is successfully removed from the crater.

The primary mission objective is to free the lander. Each competitor can choose between the following two approaches to achieving that objective:

Free the Lander from the Crater by pushing or pulling it.

or

Free the Lander by placing the Battery in the battery box and firing the Grappling Gun.

In addition, there are several, optional objectives. Competitors can tackle them or not, depending on how confident they feel:

Optional Mission Objective 1: Retrieve and place as many damaged components on the lander body as time permits; these components may be placed on the lander before or after it has left the crater.

Optional Mission Objective 2: Retrieve the damaged tire, and insert it into the lander in its original location.

Optional Mission Objective 3: Retrieve one or more of the rock samples and place in the sample box.

Competitors are free to tackle objectives in any order. Optional objectives may be completed at any time. For example, it is OK to complete an optional objective, followed by the primary objective, and then another optional one. Keep in mind, though, that optional objectives count toward scoring only if the primary objective has also been achieved; competitors failing to reach the primary objective receive no points for any of the optional ones.

Primary Mission Objective: Freeing the Lander

The Primary objective is to free the lander from the crater. The following rules pertain to this objective:

1. When the challenge starts, the grappling gun is loaded, and the lander is placed so the grappling gun is pointing to the left, as shown in Figure 9-1.

2. The rescue robot may push or pull the robot; bonus points will be awarded if the lander's hook is used.

3. If the battery is retrieved and placed in the battery box (see Figure 9-3), the grappling gun may be fired. Firing the grappling gun will award more points than pushing or pulling the lander from the crater.

Figure 9-3. The battery placed in the battery box

Optional Mission Objective 1: Retrieving Damaged Components

Five damaged components rest outside the crater: a tire, vacuum, cutter, communications umbrella, and battery. Points will be awarded for each damaged component successfully retrieved.

1. When the challenge starts, the damaged tire must be placed so the axle is pointing down, putting the damaged tire at a raised angle as shown in Figure 9-4.

2. When the challenge starts, the communication umbrella must be placed as shown in Figure 9-5; the axle post must be touching the mat in such a way that the "umbrella" (consisting of four claws) is at a raised angle.

3. The battery, vacuum, and cutter may be placed in any orientation (face up or face down, for example) in their designated location.

4. When a damaged component is retrieved, it must be placed on the lander's body for full points to be awarded. If any part of a damaged component is touching the challenge area (or mat), no points will be awarded.

5. If the lander is pushed or pulled from the crater, the battery may be placed
    anywhere on the lander's body and does not necessarily have to be placed in
    the battery box. No bonus points will be awarded if the battery is placed in the
    battery box but the lander is pushed or pulled out; the grappling gun must be
    fired for those bonus points to be awarded.

Figure 9-4. Proper placement of the damaged tire

Figure 9-5. Proper placement of the communications umbrella

Optional Mission Objective 2: Attaching the Damaged Tire

Teams may choose to place the damaged tire in its original, undamaged location on the lander. This optional mission will award additional bonus points if successful. The following conditions related to the damaged tire can affect the score:

1. When the time limit expires, if the damaged tire is inserted into its original location on the freed lander, bonus points will be awarded. If the lander is not freed but the damaged tire is inserted properly, no points will be awarded as the primary objective was not completed successfully.

2. If the damaged tire cannot be properly inserted and time permits, a team may attempt to retrieve the damage tire and place it on the lander's body.

Optional Mission Objective 3: Salvaging the Rock Samples

Teams may attempt to salvage the rock samples for bonus points. These rocks must be placed in the sample box on the rear of the lander for points to be awarded. Here are the rules related to the rock samples and their placement:

1. When the challenge begins, all 12 sample rocks must be placed in the area designated in Figure 9-1.

2. The rocks must not be touching, and all 12 must fit inside an area no larger than a 6-inch diameter circle.

3. Bonus points for rock retrieval will not be awarded unless the lander is successfully freed from the crater (that is, the primary mission objective is accomplished).

Scoring the Challenge

The challenge score will be determined using the following system:

- Primary mission objective accomplished, push or pull: 10 points
- Primary mission objective accomplished, gun fired: 20 points
- Optional mission objective 1 accomplished: 5 points for each part
- Optional mission objective 2 accomplished (tire fixed): 15 points
- Optional mission objective 3 accomplished: 2 points per rock

Earning Bonus Points

Award bonus points as follows:

- All optional mission objectives accomplished: 15 points
- All rocks retrieved and placed in the sample box: 10 points

Using the Mars Base Gamma Mission Scoring Form

Use the Mars Base Gamma Scoring Form to track the completions of the mission objectives and tally the final challenge score.

Note: Mars Base Gamma Scoring Form may be downloaded at www.marsbasecommand.com. Click the Mars Base Gammaa Scoring button.

Frequently Asked Questions

Following are some frequently asked questions, and their answers:

1. *Do I have to successfully complete any of the Optional Mission Objectives?*
   No. Only the primary mission objective must be completed in order to obtain points. If the lander is freed by pushing, pulling, or firing its grappling gun, the primary mission is successfully and no further objectives must be met.

2. *What if I successfully place the battery but cannot fire the grappling gun?*
   You will get the bonus points awarded for having a damaged component placed on the lander if the lander is removed from the crater (by pushing or pulling).

3. *Is there a height or weight limit for the rescue robot?*
   No. But you are limited to the parts you can use: one NXT robotics kit plus leftover parts from one Resource Set (if one is used).

4. *What happens if a damaged part is placed on the lander but falls off when it is removed from the crater?*
   No points will be awarded for any part that touches the mat when the time limit expires.

5. *What if I encounter a situation that's not covered by the challenge rules?*
   Make your own ruling on the matter that is in agreement by all parties participating in the challenge.

Adding Novice and Expert Rules

As an alternative to the standard mission challenge rules, some novice rules have been created that will make the challenge a little easier to successfully complete. These novice rules are as follows:

1. The challenge time limit is doubled to 8 minutes.

2. The grappling gun may be fired without the battery; the battery is not even required to be placed on the lander's body.

All the other regular rules still apply.

As an alternative to the standard Mission Challenge rules, some expert rules have been created that will make the challenge a bit more difficult to successfully complete. These expert rules are as follows:

1. The challenge time limit is cut in half to 2 minutes.

2. All damaged components must be retrieved and placed on the lander's body for any bonus points to be awarded.

Again, all the remaining rules still apply.

Summary

The Mars Base Alpha challenge is supposed to be fun. While every effort has been made to keep the rules easy to follow, it's simply impossible to predict every question that will arise regarding the rules.

Therefore, teams and individuals will be given the benefit of the doubt when it comes to interpreting the rules. If a rule is unclear to you, think about the overall objective of the missions, and make your best decision regarding how you will resolve or enforce that rule.

Mars Base Command is about learning, doing, and having fun—not spending large amounts of time worrying about rules. Try your best to solve the missions using the information provided, and if a robot finds itself in a situation that the rules don't address, don't worry about it! Make the best ruling you can and move on.

Finally, keep in mind that your robot doesn't have any limits on the number of times it can attempt missions. Continue to refine and test your robot until it is able to accomplish all the missions.

One last thing—if you don't like the rules, change them! Give yourself more time or allow third-party sensors or whatever will make the experience more enjoyable for you.

CHAPTER 10

■ ■ ■

Internal Medicine

We Need a Plan

"Dan's vitals are stable, and transferring him back here would be the ideal situation," said Doctor Kim Homansky. "But without the Surgical Assistant online, I think treating him locally is the best bet."

Lieutenant Kristie Raleigh leaned back in Commander Evans's chair and exhaled loudly. The commander had been unable to return to Mars Base Alpha since the meteor shower two days earlier; all travel between bases had been suspended until Alpha's power issues were completely resolved. Adding even one additional person to the fragile life support system before the engineers stabilized the power generation grid could cause another set of power failures. The chair would have been comfortable on any other day, but today, she was second in command and facing a handful of difficult command decisions, including how best to treat the injured commander currently stuck just five miles away at the new Mars Base Beta, which was lacking its own medical team.

"It's our only option, right now," replied Kristie. "Mars Base Beta's rover is damaged. And even if it was functional, I'd have to swap out two or three staff members from Alpha with the commander and his escorts to keep life support stable. Commander Evans is going to require on-site medical attention at Beta. So how can we make that happen?"

"I take it that sending me over there isn't an option?"

"Kim, if just Beta was damaged in the meteor shower and we were still fully operational, I'd send you over immediately. But you've got five injured in your med lab, and two of those are critical. And Dan's injuries aren't life threatening, so sending you over there isn't the logical choice."

"Beta has no trained medical staff on site, Kristie. Dan's eye needs treatment, but it's the hand that has me super concerned."

Kristie leaned forward and rested her elbows on the desk. "Yeah, I'm not up to speed on Dan's prosthesis. What's the issue there?"

Doctor Homansky placed two sheets and a couple of photos on the desk in front of Kristie. One was an engineering schematic that Kristie immediately recognized; the other was an x-ray.

When the meteor shower hit, all of the lifters at Mars Base Beta had shut down as the control room's computers lost power. Commander Evans had been operating one of the lifters to move the base's new 3-D printer system from the rover to the cargo bay. The loss of power caused the printer to spin wildly. Commander Evans had exited the lifter and tried to get control of the heavy cargo, resulting in major damage to the rover and his right hand. Part of the rover's artificial diamond windscreen had shattered, and flying fragments had cut the commander's left eye as well.

"This is the last x-ray taken of Dan's hand, from six months ago. You can see all the components that have been inserted, including replacements for the smaller bones here," she indicated with her finger, "and here. I wish they had the ability to take an x-ray at Beta, but all I've got are these digital photos they sent over. Keep in mind that the hand is still seventy percent natural. Dan's surgeon did a great job keeping most of the original bone, tendons, and muscle."

Kristie looked at the x-ray and then directly at Kim. "Sorry, Kim. I wish I had studied harder in my anatomy class."

Kim smiled and placed the schematic over the x-ray. "Let's try this. See here, near the wrist? Based on the digital photos and the fact that Dan cannot move his fingers, I believe the median nerve is being compressed. Probably from a fragment of the replaced Scaphoid or Lunate bones. That's an easy fix if I had him here. But the longer that nerve is compressed, the more likely permanent damage to the nerve will be done."

"I hate to ask, but is there any way one of the techs over there could make an incision and fix this?"

"Maybe, but I wouldn't recommend it. I'm just guessing at the cause. But the compressed nerve isn't the biggest problem. See this?"

Kim was pointing at a small square object embedded just below the wrist. Kristie turned her head and read the text on the schematic.

"Electrolytic gel power supply? Those are fairly standard aren't they? The body handles the recharging with its own electrolytes, and the power is used to assist the microservos with movement."

"That's the textbook engineering definition," replied Kim. "But based on the coloration of the wrist, the swelling, and the elevated heart rate, I think there's been a fracture in the gel pack. Dan's body appears to be having an adverse reaction to a leak in the battery."

Kristie leaned back in the chair again and closed her eyes. "Life threatening?"

"No, fortunately. But the leak is going to continue to enflame the nerves and muscle tissue. I can send over some premixed injections that can slow down the reaction, but the longer we go without fixing the leak, the more likely he'll lose the entire hand. The eye can wait; it appears to be just superficial cuts that I can treat later."

"We need a plan," said Kristie.

It Can't Be That Simple

Mars Base Alpha, Section C, Medical
August 22, 2062 at 9:17 AM (Greenwich Mean Time)

Lieutenant Raleigh and Doctor Homansky stood facing the damaged Surgical Assistant (SA). The term "assistant" would have been humorous on any other day, as the robot was fully capable of handling hundreds of surgical procedures on its own—no human surgeon required. If the SA were operational, Kristie would have had it placed in the back of one of the working rovers and driven over to Mars Base Beta immediately. But as with many of the facilities and much of the equipment in Alpha, the power outage had damaged the Surgical Assistant, and the engineering team had not yet been able to find and fix the problem.

"I could be at Beta, make the repairs to his hand, and be back here with him in less than twelve hours," said Kim. "Three of my patients are stable, and Matthews in ICU is starting to show signs of stabilizing."

"Still too big a risk," said Kristie. "Mars Base Command has already been informed of the situation and given me direct orders to keep you here."

Kim sighed. "I don't know what to do here, Kristie. I'm the only person qualified to do the repair. I just can't believe that we don't have a backup SA."

"Too expensive," replied Kristie. "At least according to Command. I'm going to push hard for a spare, but that's not going to help our current situation."

Kristie stared at the SA and the numerous scalpels, lasers, and other tools that made up what most engineers called "the business end" of the device. Kristie knew robots but not those that performed delicate medical procedures. Her background included programming the maintenance robots that performed welding and cutting and inserting replacement parts on the base's critical machinery used for day-to-day operations.

"Wait a second," said Kristie. "It can't be that simple."

Kim turned her head. "You have an idea?"

Kristie nodded.

"I do. But you're not going to like it."

Kristie smiled at the doctor and motioned her to follow.

The Tools are the Same

Mars Base Alpha, Section D, Control Center
August 22, 2062 at 9:32 AM (Greenwich Mean Time)

"No way! You are not going near my patient with one of those things!" yelled Kim.

Brian Platt, one of Alpha's engineers, was laughing. "First, you send a robot to fix our power problems, and now, you want to cut on Commander Evans with one?"

Kristie wasn't smiling. Commander Evans needed immediate medical treatment if he was to keep his hand, and she was going to do everything in her power to help him.

"What's the difference?" asked Kristie. "Your SA is a robot. This is a robot."

Kristie had placed one of the engineering diagnostic and repair robots on the small table in the control room. The size of shoebox, it was one of the most widely used robots on the base with its graspers, cutters, and lasers. Most engineers used them to assist with soldering and replacing components inside the many computers and machines scattered around the base, but Kristie was looking at the device in a completely different way now.

"My Surgical Assistant is designed for medical procedures. That… thing… is used to replace conduit wire in the walls!" replied Kim.

Kristie turned to Brian. "All of the automated systems and equipment in the base use the same type of control circuitry, right?"

Brian nodded. "Yeah, but the programming is different for each device."

"Exactly," said Kim. "Cutting out a damaged LCD screen isn't the same as cutting a piece of bone, Kristie."

Kristie smiled. "You're, right, but the tools are the same. Look here." Kristie pointed at the diagnostic robot's tools. "Small cutters, two lasers, a pair of graspers that are no bigger than the ones I saw on the SA."

Brian looked closely at the small collection of tools. "Huh, you're right."

"No! You just can't program this little thing to cut open the commander's hand and start messing around with bones, nerves, and muscle. It's not programmed with the proper procedures."

Kristie grabbed Kim's shoulders and stared her in the eyes. "Not yet."

Are you Serious?

Mars Base Beta, Recreation Area
August 22, 2062 at 12:45 AM (Greenwich Mean Time)

The video feed from Beta was still grainy, but Power Systems Engineer Dale Richards's voice was crystal clear.

"Are you serious?" he asked, looking over his shoulder at the diagnostic robot sitting on a table behind him.

"It will work, Dale," said Kristie. "And Doctor Homansky will be here with you during the entire procedure."

Kristie looked back at Kim who stood behind her but still within the camera's range.

"Yeah, she looks real confident," said Dale.

"Tell him, Brian," said Kristie.

Brian stepped into the camera's view and nodded. "We took the Surgical Assistant's latest firmware version and flashed the diagnostic robot's control chip. You won't have the SA's voice control, but it will wirelessly display all the relevant data to any computer screen you want to set up there as well as send the data to Doctor Homansky's screen over here."

Dale shook his head. "And this other stuff?"

"That's some injections I'll need you to give Commander Evan's before we begin as well as the replacement components for his hand," said Kim.

"Shots give me the heebie-jeebies," said Dale with a shudder.

"You'll do fine," said Kristie. "Doctor Homansky will be right here with you."

"You mean right *there* with *you*," said Dale.

Kristie laughed. "Just let the robot do the work, Dale."

"You're the boss," replied Dale.

"Hey!" A voice suddenly interrupted from off camera. "I'd like to speak to *the boss*."

Kristie heard the distinctive Texan drawl in the background as Dale rotated the camera. Sitting on a chair with his right arm elevated, Commander Evans stared back, his left eye covered.

"I thought you were sleeping," said Kristie.

"Tried to but now that I see that little diagnostic robot…"

"That little robot is going to save your hand, Commander," said Kristie. "Are you in pain?"

"My eye hurts a bit, but the hand is just numb."

"I'm still going to have Dale give you a sedative, Commander," replied Kim. "I have something else if you'd like to be put under."

"No, thanks," replied the commander. "I think I'll watch."

Dale shook his head. "Can I be put under?"

Kim smiled.

"Is this going to work, Doc?" asked Commander Evans.

Kim looked at Kristie and then back at the camera. "I think so, Commander. I'm not happy about it, but the procedure is fairly simple—a few parts moved or replaced and we're done. The little device will even patch it all up with some spray-on stitches."

"Well, let's get moving then," said Evans. "I've got a lot of work to do."

Kristie stepped back from the desk. "Commander, I'm going to let Kim and Dale take the lead here. I think you're in good, uh, hands."

Brian shook his head. "Never a dull day at Alpha," he said.

Kim nodded at Dale. "OK, first, I need you to remove the items I packed in box A."

Your Turn

Mars Base Beta
August 22, 2062

Was Commander Evans's hand successfully repaired? You're about to find out. Continue with Chapters 11 and 12 and assemble the mechanisms that will be used in the challenge. After building these devices, you'll next need to construct and program a robot to successfully repair Commander Evans's hand following the rules and challenge setup found in Chapter 13.

Good luck!

CHAPTER 11

Getting Started with Commander Evans's Hand Challenge

To successfully complete the Mars Base Beta challenge (challenge number 3), you will be required to construct a diagnostic robot that will inspect and repair the damage done to Commander Evans's prosthetic hand. Thus, you must also build that hand.

The prosthetic hand received crushing damage, causing the wrist actuator's battery to begin leaking the electrolytic gel into the surrounding tissue. To simulate the hand and the areas that will need to be repaired, you will need to assemble the models found in this chapter and in Chapter 12. Figure 11-1 shows the finished collection of models that you will be building.

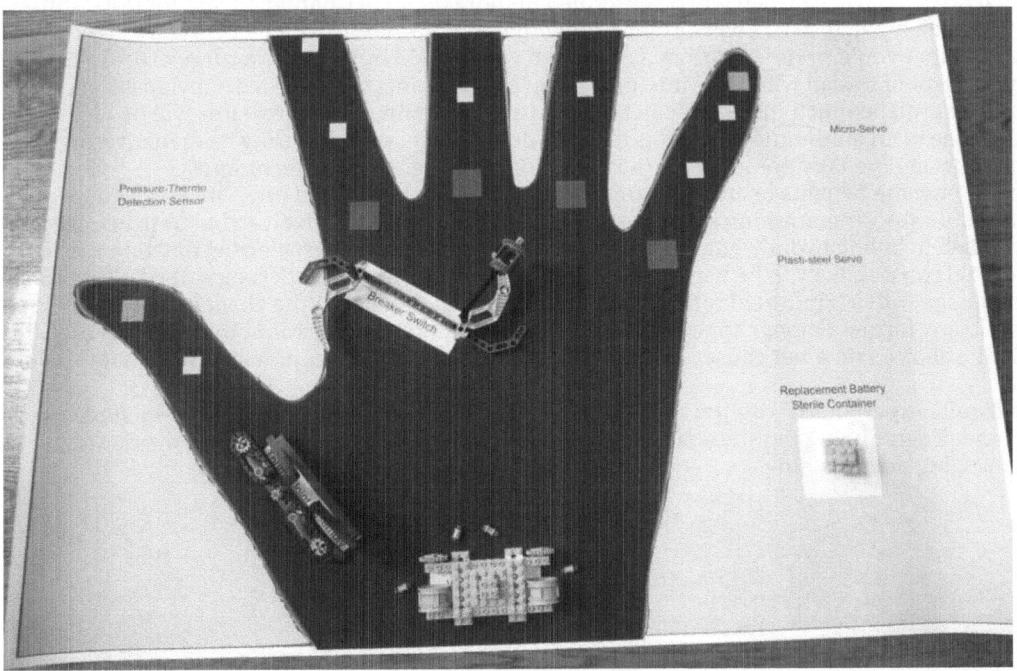

Figure 11-1. The components in Commander Evan's hand that must be repaired

The three models to be built include the wrist actuator (with electrolytic gel battery inserted), the index finger breaker switch, and the thumb joint assembly. In addition to performing certain actions on these models, you will also be tasked with removing as much of the leaked battery gel as possible.

Tackling the Prosthetic Hand Challenge

You'll first want to assemble all of the models using the build instructions found at the end of this chapter and throughout Chapter 12. Chapter 13 will provide you with instructions for setting up the challenge using either the challenge mat or your own challenge area.

After you've assembled the prosthetic hand components, you'll want to examine how each of them works and begin brainstorming how your diagnostic robot might achieve the desired actions. These actions include

1. In the wrist actuator, remove the damaged battery and replace with working battery.

2. Reposition the index finger breaker switch by rotate the switch's arms so they are pointing toward one another and sliding the breaker from one end to the other to reset.

3. Repair the thumb joint assembly by rotating the orange connectors to point outward and rocking the assembly left or right to test joint functionality.

The actual tasks to be performed will be described in more detail in Chapter 13, but for now, you really need to decide is whether you will design a robot that is autonomous and can handle the repairs on its own or whether you'd prefer to design a robot that can be held by you (like a surgical tool). The difference in your robot design will come into play when bonus points are awarded. Obviously, an autonomous robot will be much more difficult to design and program, but the rewards will be higher as well. The downside to an autonomous robot is that any damage that occurs during its maneuvering cannot be undone, and penalty points will be awarded for improper surgical techniques.

Be sure to weigh the benefits of an autonomous robot versus a hand-held one. Obviously there are points involved, but the programming aspects of an autonomous robot are likely to be more involved. Challenge yourself to build beyond your comfort level however. If you are just starting out with the NXT kit, the hand-held robot is likely to be a much easier and far less frustrating experience. It is fully possible to win the challenge with a hand-held robot that completes all challenges in the challenge time provided and does not do any further damage to the prosthetic hand. An autonomous robot that completes all challenges in the allotted time but causes damage will still end up with fewer points than its hand-held counterpart.

If you are quite experienced with the NXT kit, give the autonomous robot a try—you will find that the building and programming requirements will test your abilities and push you to discover new methods for tackling the problem.

Examining the Challenge Area

Because your robot (either autonomous or hand-held) can move anywhere it likes, you will not be limited to the boundaries of the challenge area. The area you will be working with is the commander's hand, and because of this, the models must be secured in place (with tape or Velcro). Moving any of the mechanisms will be treated as additional damage to the hand and penalty points will be received.

Because avoiding further damage to the hand is so important, you will be allowed to place colored tape anywhere in the challenge area to assist your robot with maneuvering (using the Light or Color sensors, for example). You may also play additional parts (from the Educational Resource Set or any remaining parts from your robot kit) outside the line boundary that define the hand's shape. These could possibly be used by the Ultrasonic or Touch sensors as methods for determining location and/or distance. You may also designate the robot's starting area, and no penalties will be assessed for touching your robot (if it's autonomous) and bringing it back to the starting area. As long as you pick up the robot and move it without further damage to the hand, no penalty points will be received for touching the robot before the challenge time expires.

Creating the Wrist Actuator and Simulating Battery Gel Leaks

The first half of the building instructions for the prosthetic hand challenge are provided in the remainder of this chapter. Each image shows the pieces you will need to locate in the LEGO Education Resource Set and their quantities. You will also see where these pieces are to be placed.

Chapter 12 will continue with the building instructions for the remaining models: the Index finger breaker switch and the thumb joint assembly. You'll take your assembled models and proceed to Chapter 13 to set up and run the challenge.

Note   If you are uncertain about the placement of a piece in a figure, jump ahead to the next image or even go back to a previous image to make certain you've built the model correctly so far. Examine the figures carefully, and you should be able to correctly identify the pieces and their final location on the model.

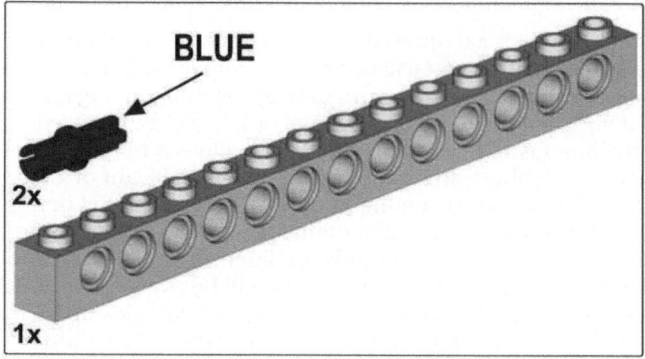

1

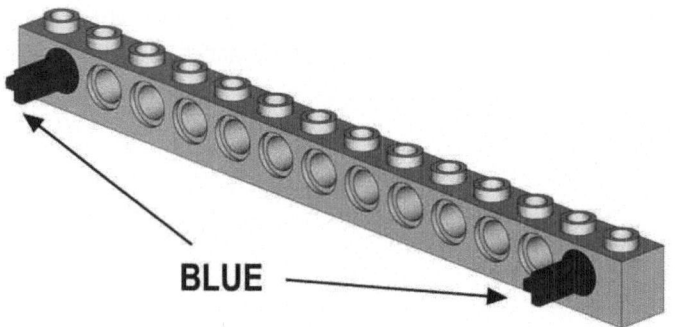

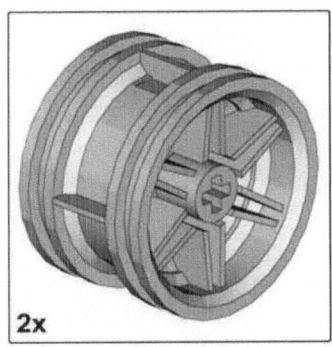

2

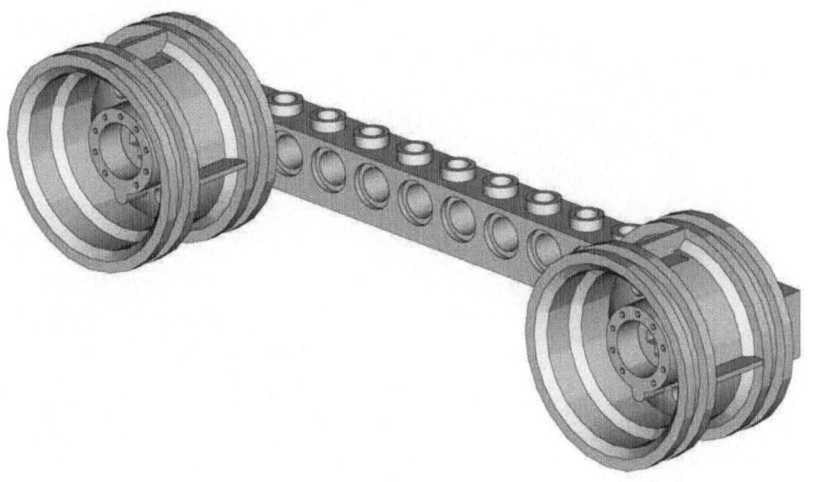

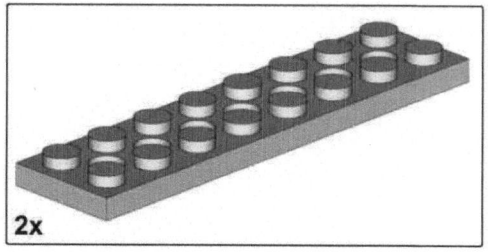

3

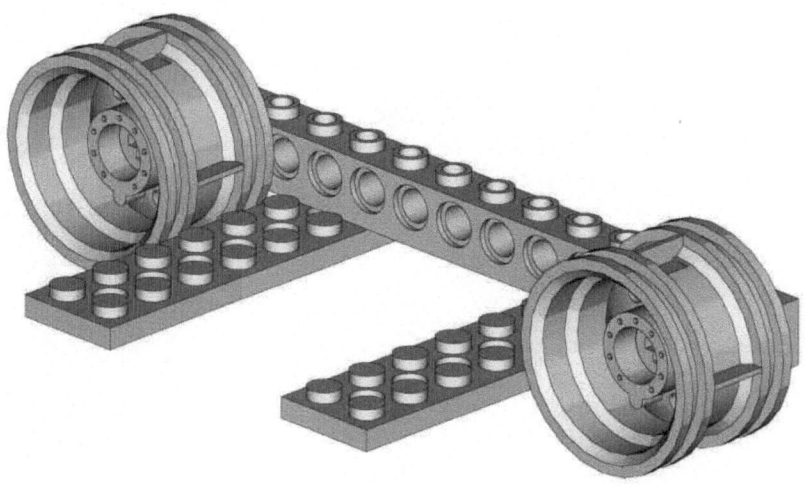

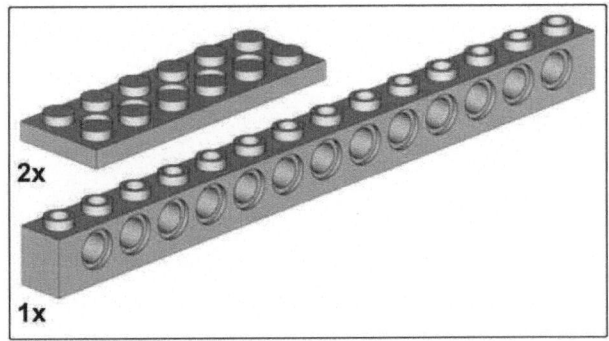

4

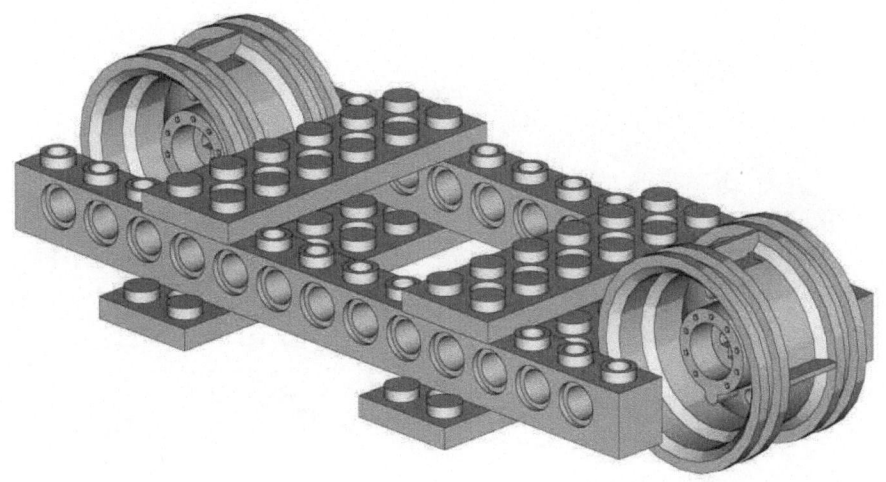

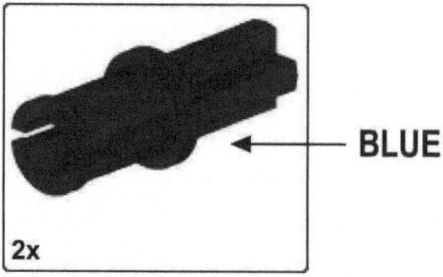

BLUE

2x

5

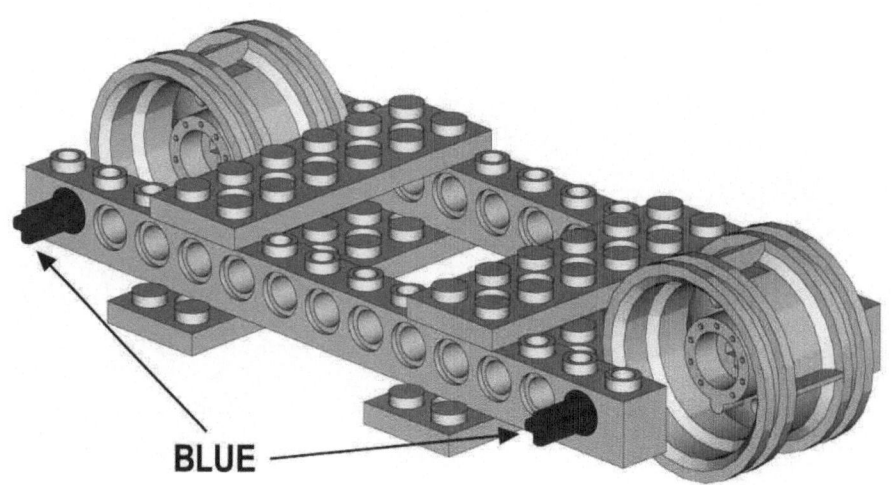

BLUE

BLUE

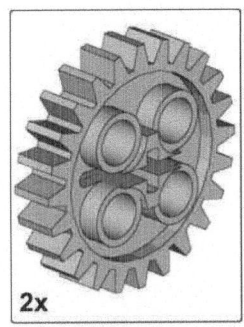

6

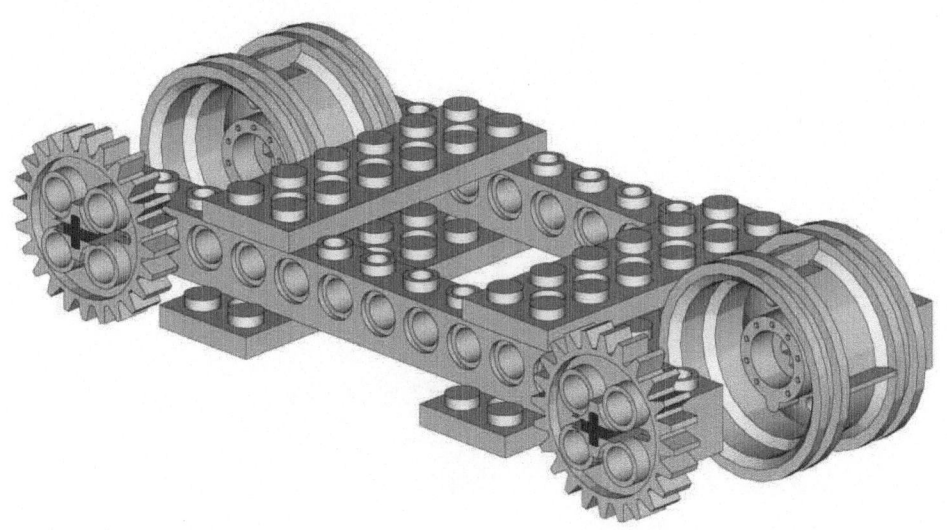

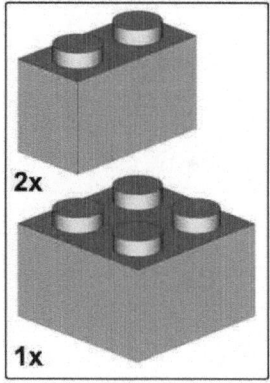

7

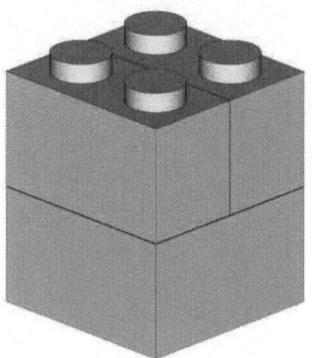

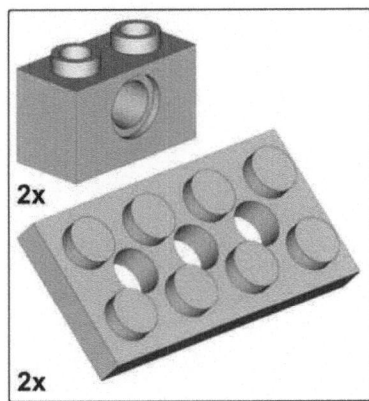

8

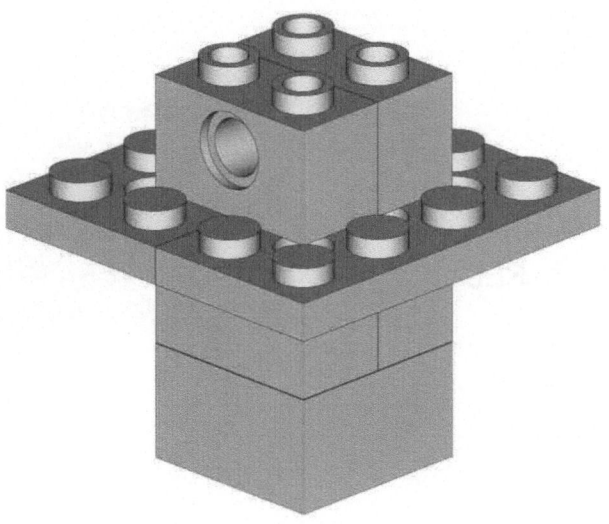

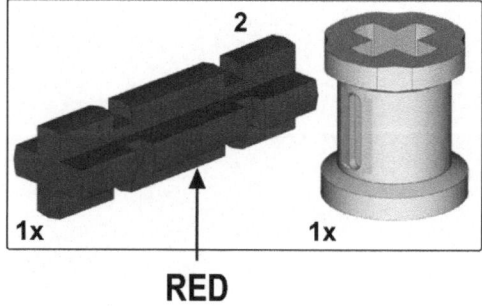

RED

9

RED

Summary

You've now got half of the models required to run the third challenge. Continue reading Chapter 12 for instructions on finishing up the remaining models.

CHAPTER 12

■ ■ ■

Finishing Commander Evans's Hand

You'll find the building instructions for the remaining two models used in the third challenge in this chapter. Each image shows the pieces you will need to locate in the LEGO Education Resource Set and their quantities. You will also see where these pieces are to be placed.

After you finish building the models, you'll need to read Chapter 13 regarding the rules of the challenge and how to set up the challenge area.

Note If you are uncertain about the placement of a piece in a figure, jump ahead to the next image or even go back to a previous image to make certain you've built the model correctly so far. Examine the figures carefully, and you should be able to correctly identify the pieces and their final location on the model.

Building the Index Finger Breaker Switch

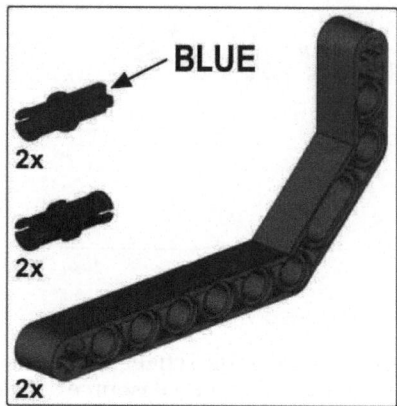

1

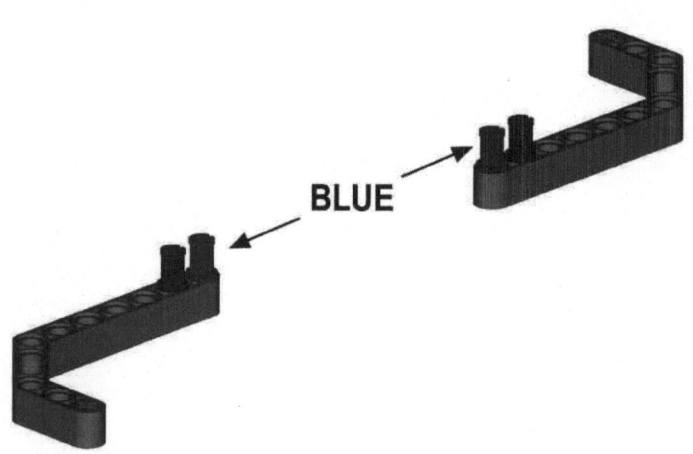

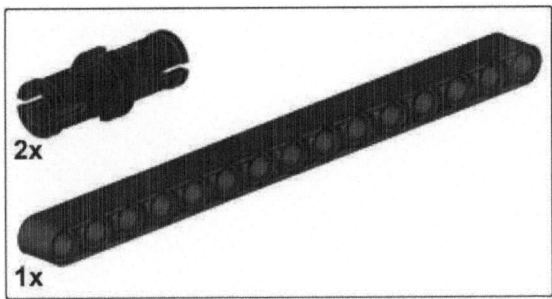

2

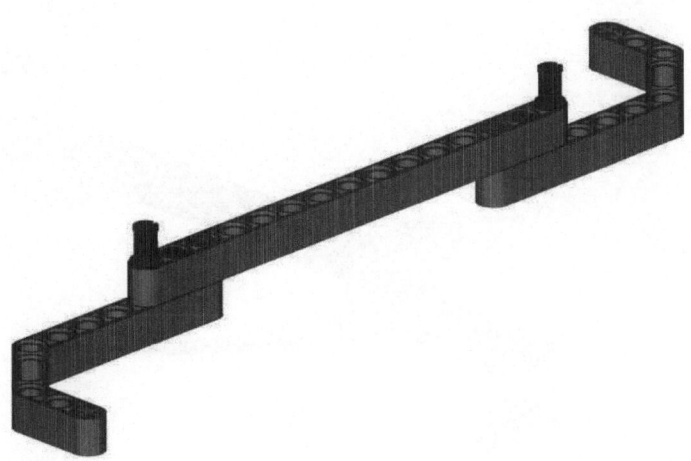

2x

3

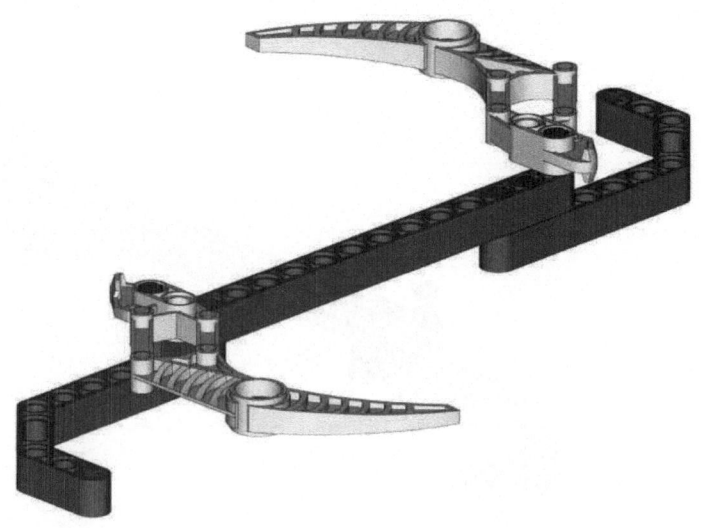

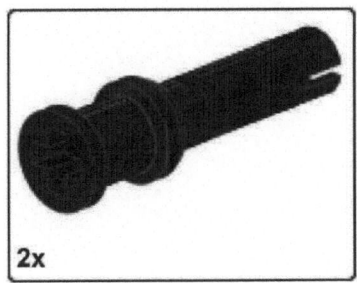

4

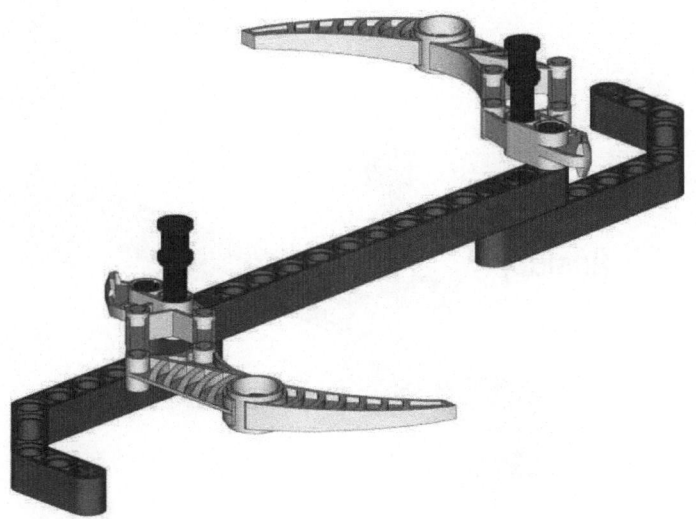

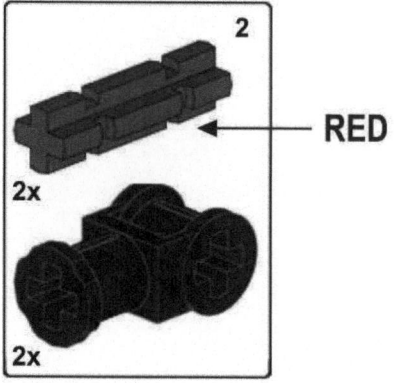

RED

2x

2x

5

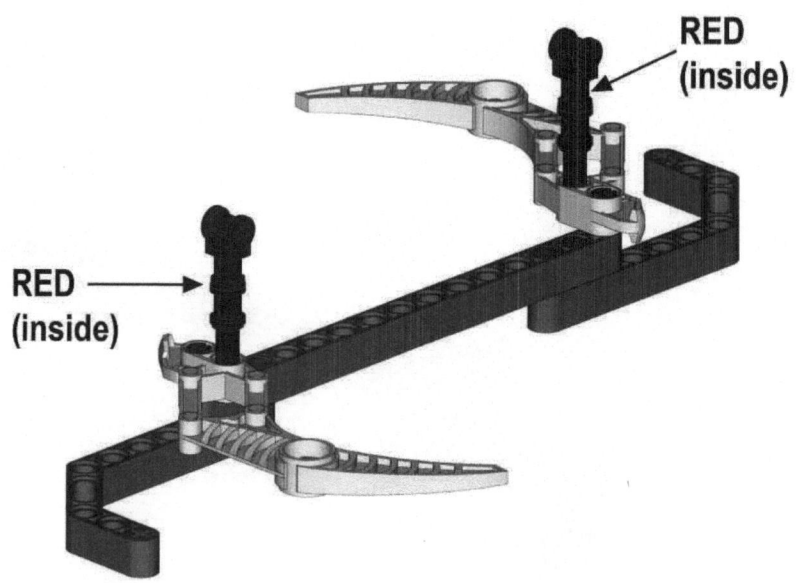

RED
(inside)

RED
(inside)

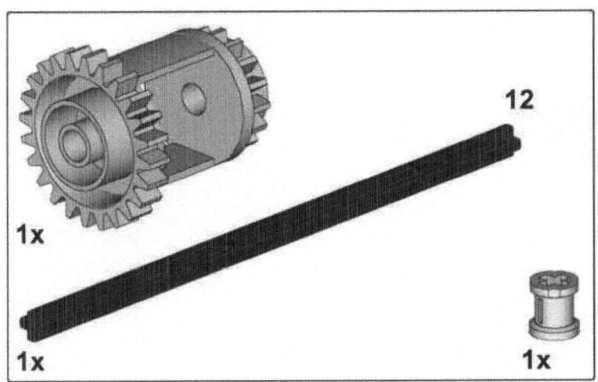

6

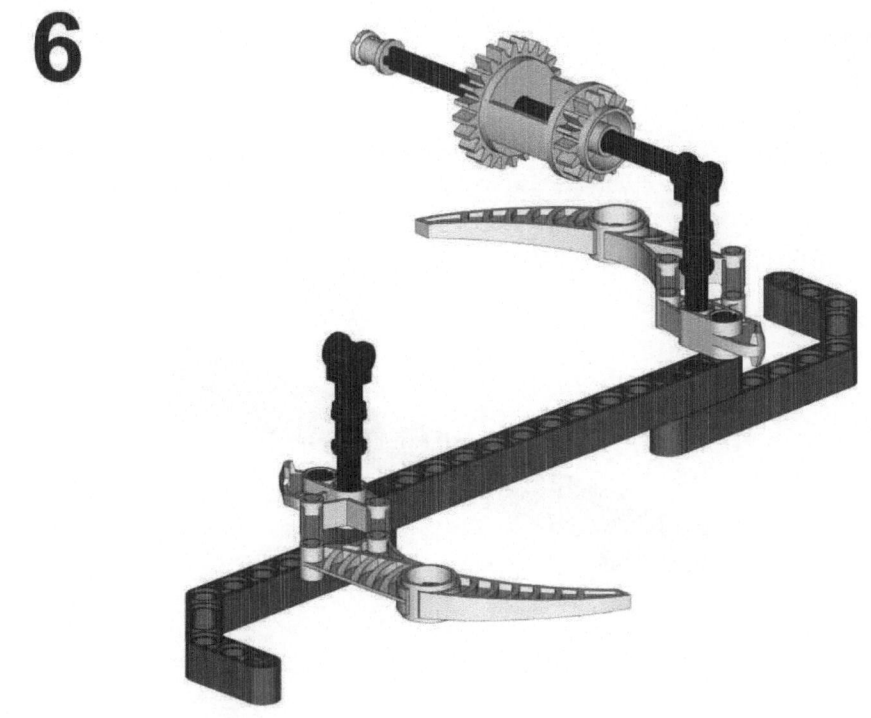

Building the Thumb Joint Assembly

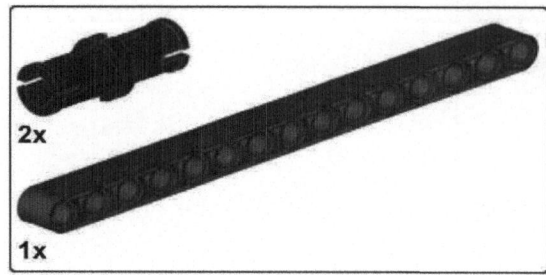

7

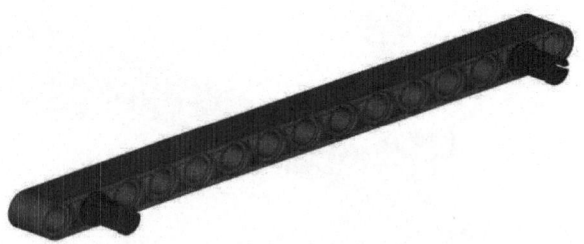

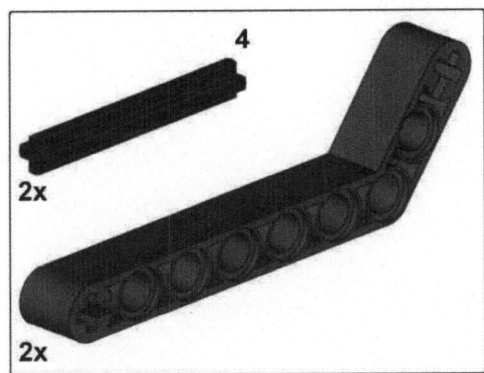

8

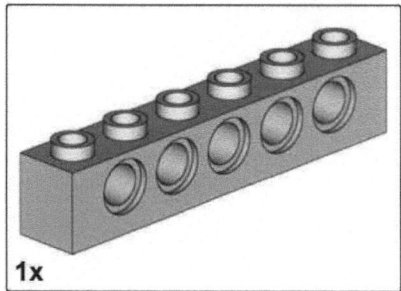

9

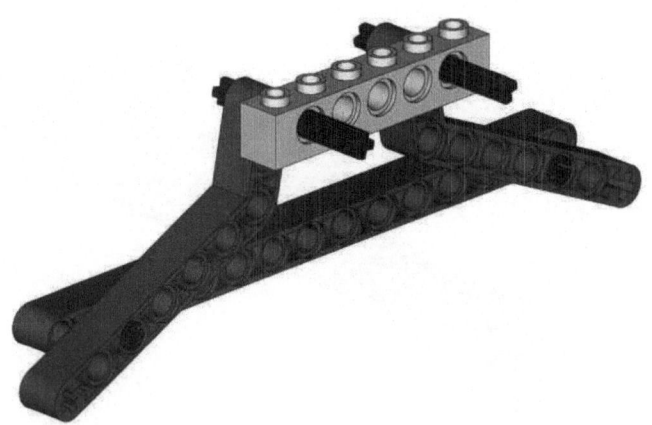

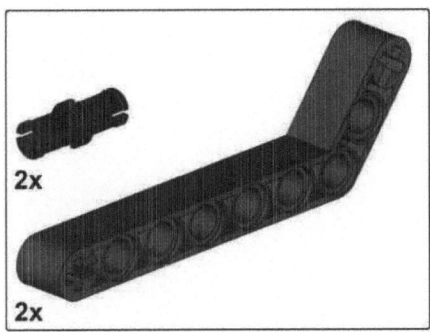

10

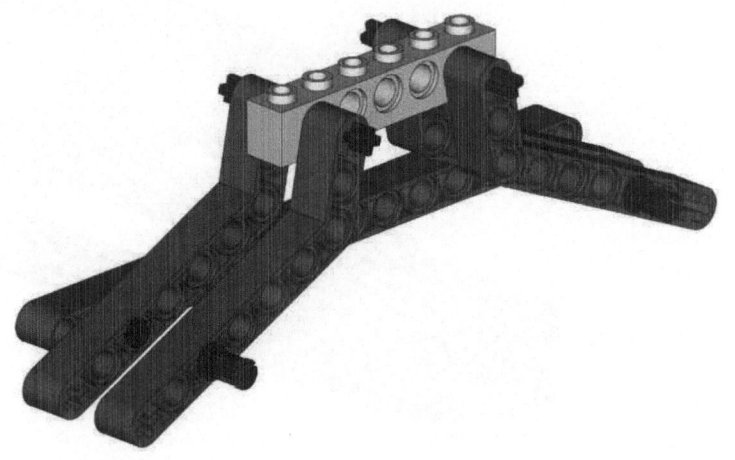

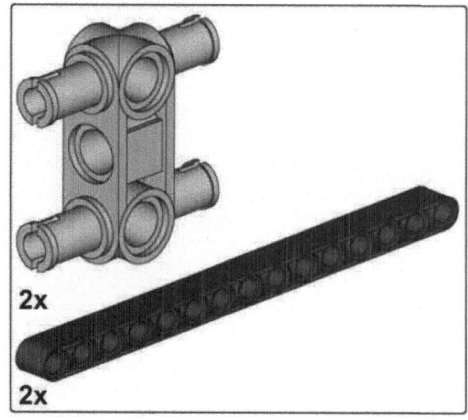

11

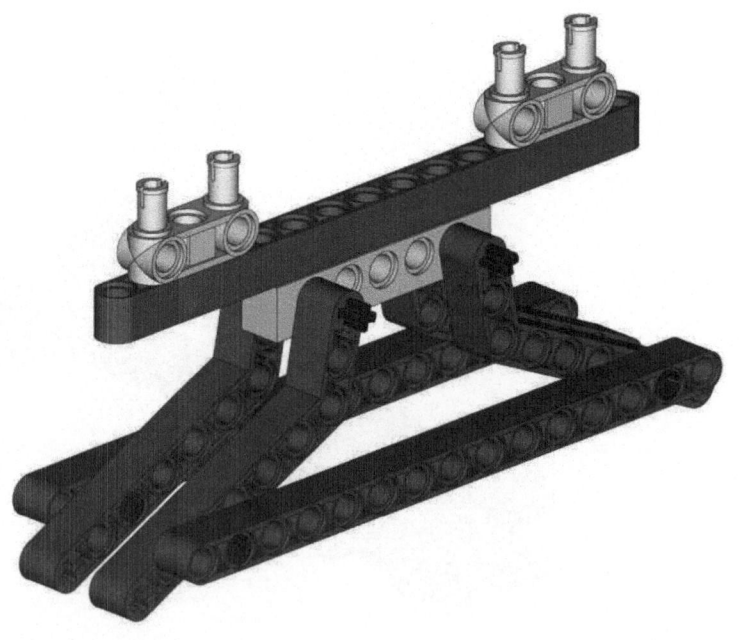

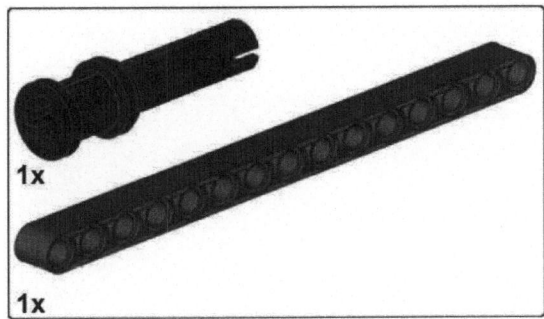

1x

1x

12

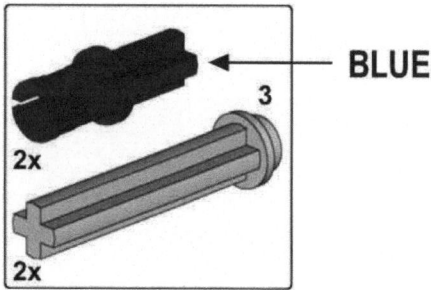

BLUE

2x

3

2x

13

BLUE

BLUE

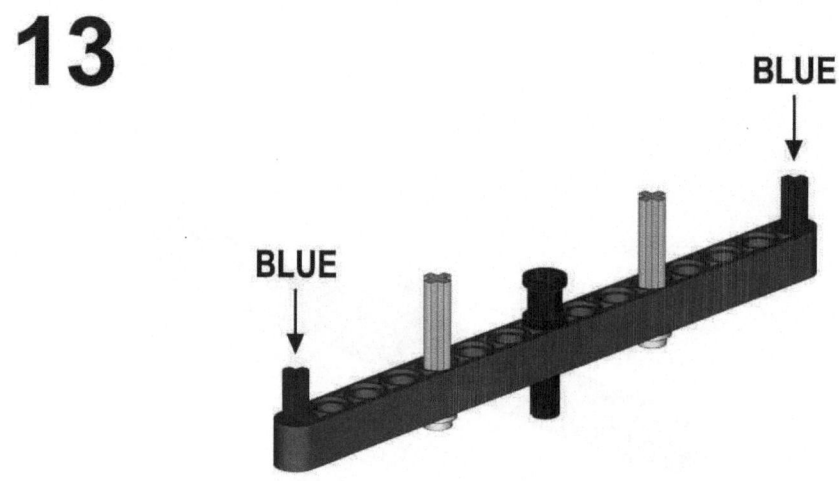

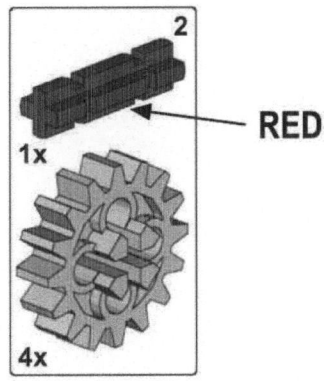

RED

14

RED

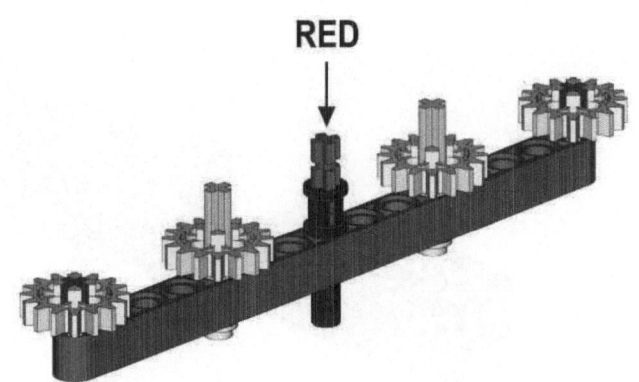

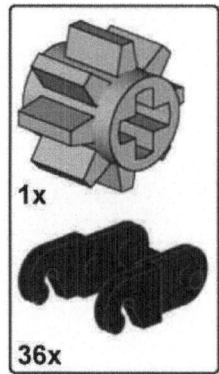

15

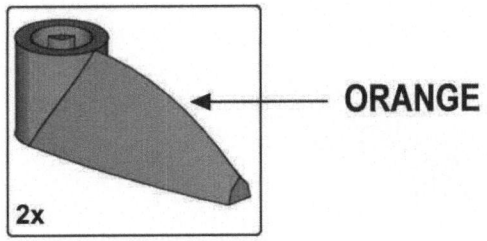

ORANGE

2x

16

ORANGE

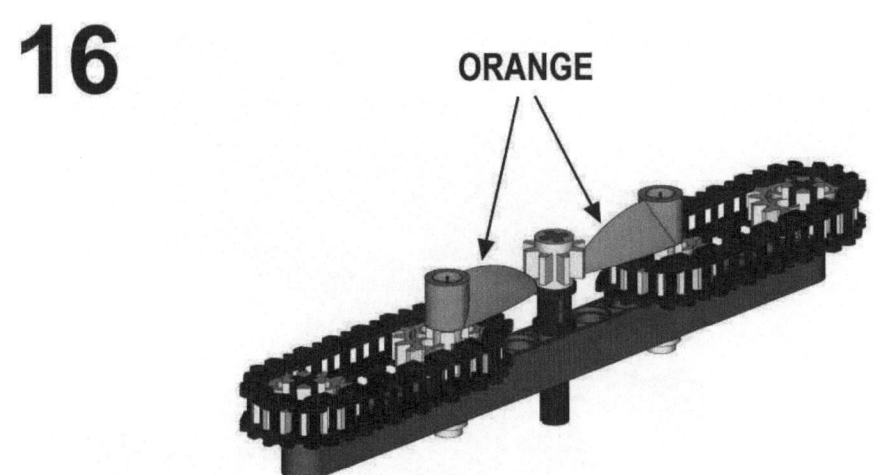

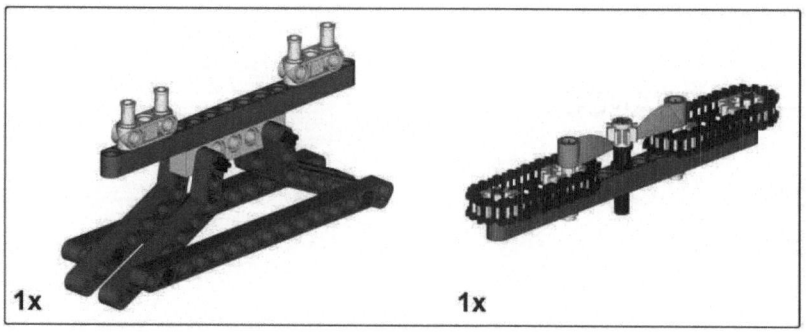

1x   1x

17

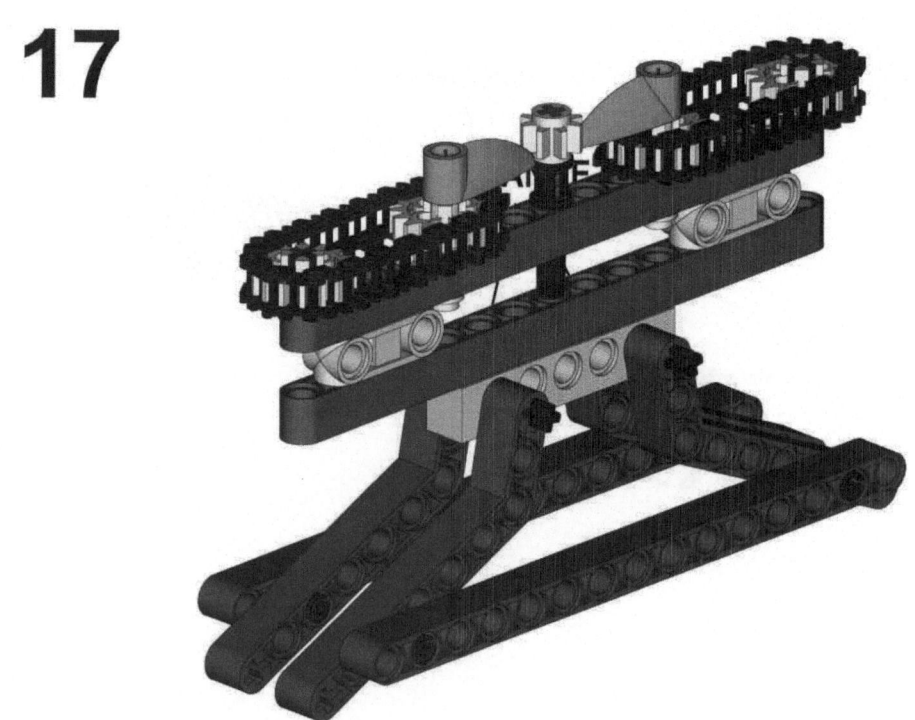

Summary

You now have all the models you need to simulate the prosthetic hand. Chapter 13 will give you the rules and show you where to properly place the models to run the challenge.

CHAPTER 13

■ ■ ■

Synopsis and Rules of the Internal Medicine Challenge

The Mars Base Beta: Internal Medicine challenge will once again test your robot building and programming skills as you design and program a robot to repair damage done to Commander Evans's right hand. The prosthetic hand has many internal components such as a battery, fingers with micro servos, an extremely complex thumb joint device, and various breaker switches that shut down power to the fingers when extreme damage is detected.

Teams or individuals will build and program a LEGO MINDSTORMS NXT robot that will either perform the repairs autonomously or as a functioning hand-held medical device. Four unique challenges must be completed to successfully repair the hand, and players should try to avoid further damage to the hand at all costs.

To successfully complete the Mars Base Beta challenge, your robot will need to perform the following actions:

1. Replace the damaged electrolytic gel battery by removing it from the wrist actuator and inserting a replacement battery.

2. Reset the thumb joint assembly, and then test its movement.

3. Reset the index finger breaker switch.

4. Remove battery gel spillage from the area surrounding the wrist actuator.

Points will be awarded for each repair that is successfully made, and points will be lost for additional damage done to the hand during the surgical procedure (more details on the rules and points will be provided at the end of this chapter).

Setting Up the Challenge Area

The challenge area (CA) that defines the boundaries for your robot must adhere to the following restrictions. Use Figure 13-1 as a guide for proper placement of the models, or download and print the Challenge Area Mat PDF file.

Note   You may download the full-color PDF and black and white PDF mat files for use in this challenge by visiting www.marsbasecommand.com and clicking the Mars Base Beta Mat button. The files may be taken to a printer and printed as a mat (color or black and white) to be used for the challenge area. Permission to download and use the file for private use is granted. The file may not be sold, and printed mats may not be sold. *The mat is not required to run the challenge.*

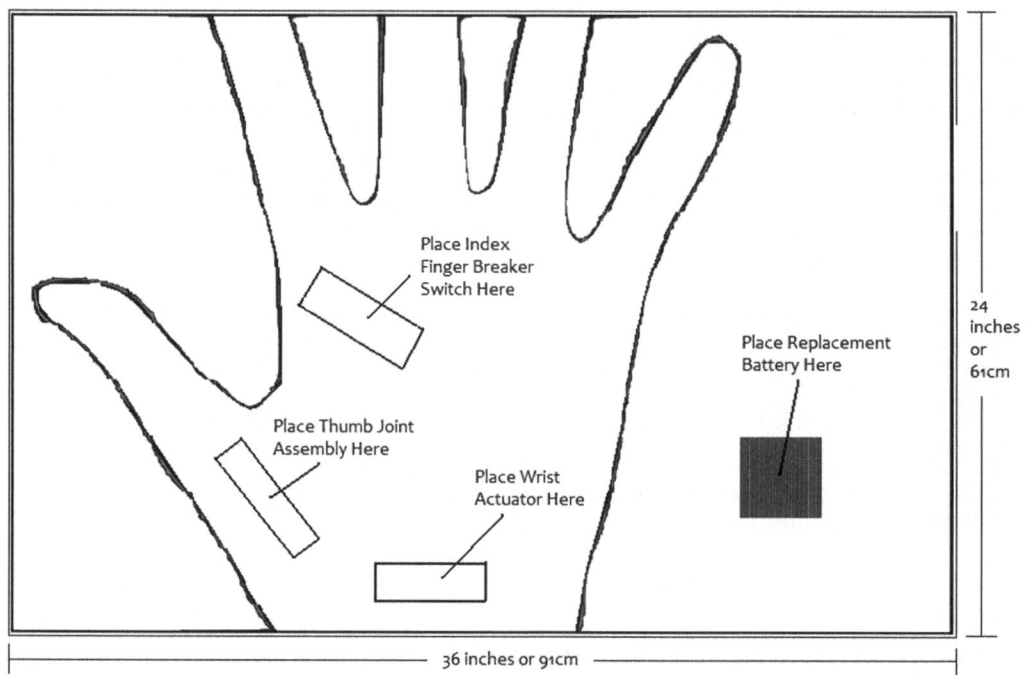

Figure 13-1. The challenge area for the Mars Base Beta challenge

The CA must be a flat surface area with dimensions of 24 × 36 inches (2 × 3 feet) or approximately 61 × 91 centimeters (refer to Figure 13-1). The greater distance is the CA width.

For the purposes of this challenge, the border of the CA is arbitrary; your robot may travel beyond the edges of the CA without any penalty. You may choose to enlarge the challenge area if you wish, using tape or any other items. I suggest that you use tape to define the outline of the hand shown in Figure 13-1, but doing so is not required.

Place the wrist actuator, index finger breaker switch, and thumb joint assembly in their proper locations as defined in Figure 13-1 or in the appropriate locations specified on the mat. The battery gel spillage (four pieces) must be placed within 1 inch (2.5 cm) of the wrist actuator but must not touch the model.

Figure 13-2 shows a photograph of the models and other components placed on the printed Challenge Area Mat.

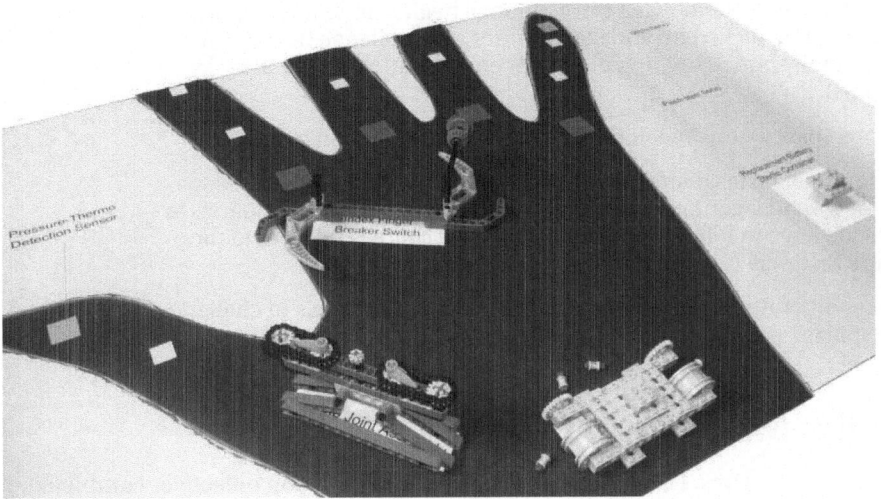

Figure 13-2. The damaged hand's components placed on the mat

Understanding the Challenge Rules

The Mars Base Beta challenge has a number of rules that must be followed by teams and individuals. If multiple teams are running the challenge, rules can be altered or removed if all players are in agreement. The rules for the Mars Base Beta challenge follow:

1. The robot created to attempt the missions must be built using parts from a LEGO MINDSTORMS NXT robotics kit (Retail or Education version 1.0 or higher).

2. A minimum of one robotics kit must be used to build the robot; special commendations are available for successful completion of missions using a single robotics kit.

3. A maximum of two robotics kits may be used to build the robot.

4. If a LEGO MINDSTORMS NXT Education Resource Set is used to build the mission models, any parts remaining in the Resource Set may also be used in the robot design.

5. Only NXT motors may be used in the robot's design.

6. Approved sensors are the Ultrasonic, Sound, Touch, Color and Light (NXT versions only).

7. The challenge time limit is 3 minutes. The challenge ends when the time limit expires.

8.   An autonomous robot must be placed outside the mat/challenge area, and its starting position can be agreed on by all challenge participants.

9.   A hand-held robot can be held when the challenge begins, or teams may agree that the robot must be picked up and started after the challenge begins. The hand-held robot may not be holding the replacement battery when the challenge begins.

10. Points are scored based only for successfully completed mission objectives.

11. Tape (any color) may be used and placed anywhere on the challenge area to assist the robot with navigation. It may be used to define the CA boundaries as well as to provide lines to follow or points on the CA. There is no limit to the amount of tape that may be used.

12. Tape, Velcro, or other items may not be used to secure the models to challenge area. The gel spillage may not be attached to the challenge area.

Understanding the Mission Objectives

This mission has four primary objectives. Points will be awarded for any mission objectives completed, and bonus points will be awarded if all mission objectives are completed successfully. Mission objectives may be attempted in any order. However, the wrist actuator mission will be considered a success only if the replacement battery is placed before time expires. If the damaged battery is removed but the replacement battery is not placed before time expires, no points will be awarded for the wrist actuator mission.

Mission objective 1—wrist actuator repair. Remove the damaged battery without disturbing the rest of the wrist actuator model. After the damaged battery is removed, insert a replacement battery without disturbing the wrist actuator model.

Mission objective 2—index finger breaker switch closure. Close the open breaker switch, and then slide the circuit connector from one side to the other. When the challenge begins, the breaker switch is open (claws pointed in opposite directions and perpendicular to base of model).

Mission objective 3—wrist joint assembly reset. Rotate the power connectors (orange pieces), so they are pointing toward each other. When the challenge begins, the power connectors are pointing away from one other, toward ends of model. After rotating the power connectors, tip the model left or right to test the thumb joint's movement.

Mission objective 4—battery gel spillage removal. Remove all four incidents of gel spillage from the challenge area without disturbing any of the mission models.

Mission Objective 1: Repairing the Wrist Actuator

The wrist actuator objective will require your robot to remove the damaged battery shown in Figure 13-3 and replace it with the working one. Use the following list to set up the wrist actuator objective as well as determine when it has been successfully completed.

1. When the Challenge starts, the damaged battery must be inserted into the wrist actuator model.

2. The diagnostic robot must remove the damaged battery without disturbing the wrist actuator. No additional parts must be removed, and no major movement to the wrist actuator must occur (the amount of allowable movement must be agreed on by all challenge participants).

3. The damaged battery must be removed from the defined challenge area before the replacement battery is inserted.

4. During the insertion of the replacement battery, the wrist actuator model must not incur any major damage or movement.

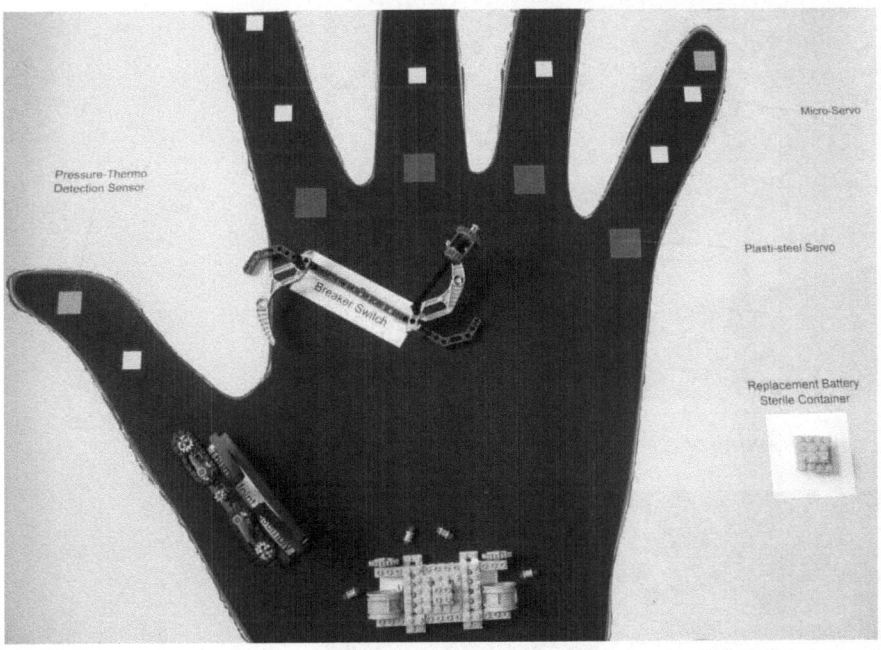

Figure 13-3. The damaged and replacement batteries

Mission Objective 2: Closing the Index Finger Breaker Switch

When the challenge begins, the index finger breaker switch (IFBS) must be in the open position. This requires that the claws be pointing in opposite directions but perpendicular to the base (see Figure 13-4). The following list will provide you with the starting and ending positions of the IFBS to help you determine if the objective has been successfully completed:

1. When the challenge starts, position the claws to be pointing in opposite directions and perpendicular to the base, as shown in Figure 13-4.

2. The circuit connector (a round component on axle) must be positioned at one end of the axle (either end is acceptable).

3. The robot must close the claws so they run parallel to the base, as shown in Figure 13-5.

4. The robot must slide the circuit connector to the opposite end of the axle.

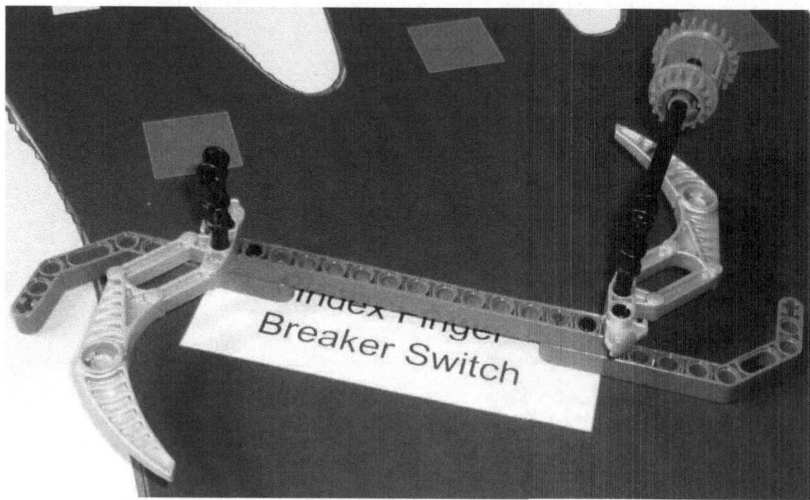

Figure 13-4. Proper starting positioning of the index finger breaker switch model

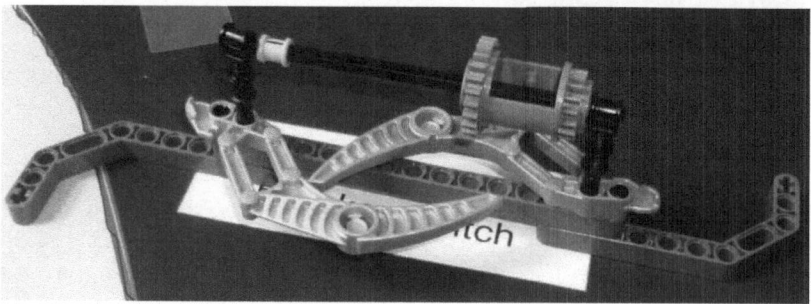

Figure 13-5. Successful completion of the IFBS mission objective

Mission Objective 3: Resetting the Thumb Joint Assembly

The thumb joint assembly reset (TJAR) objective will require the robot to perform two tasks to successfully complete the mission. The following list will explain the starting position of the thumb joint assembly reset objective as well as the two tasks that must be performed on the model for a successful completion of the objective:

1. When the challenge begins, the TJAR must have the power connectors (the two orange components seen in Figure 13-6) pointed away from each other and parallel to the model's base. The model must also be parallel to the challenge area surface.

2. The robot must rotate the orange pieces so they are facing one another (see Figure 13-7).

3. The robot must test the repaired thumb joint by rocking the model left or right (see Figure 13-7).

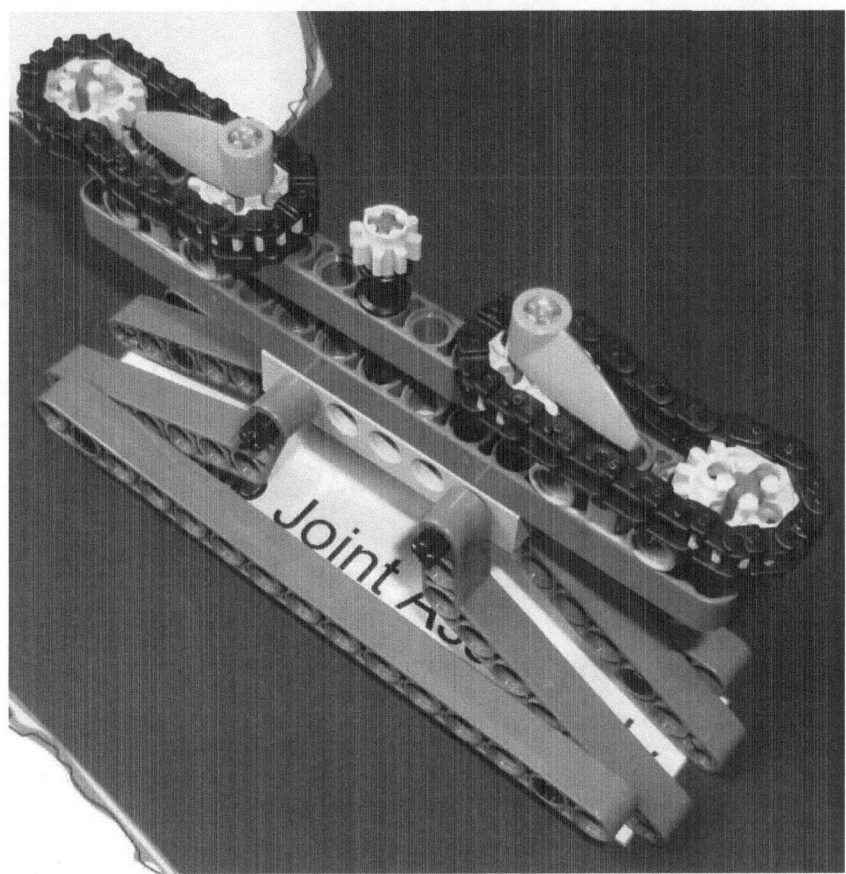

Figure 13-6. Starting positioning of the TJAR objective

Figure 13-7. Successful completion of the TJAR mission objective

Mission Objective 4: Removing the Battery Gel Spillage

The robot must remove the four instances of battery gel spillage near the wrist actuator. Keep in mind that the spillage represent toxins that have been released into Commander Evans's hand, and every effort should be made to successfully remove all spillage. Note that a high number of bonus points are available if all spillage is removed, so spend some time developing and testing a good method for removing the battery gel using your robot.

1. When the challenge begins, the four pieces representing the spilled battery gel must be placed within 1 inch (2.5 centimeters) of the wrist actuator. Players may place the pieces together on the top, bottom, left, or right, or the pieces may be divided in any manner and placed in locations agreed on by all challenge participants. Figure 13-8 shows one possible location for the pieces.

2. All battery gel spillage components must be removed from the challenge area for complete points to be awarded. Partial points are not awarded for partial spillage removal.

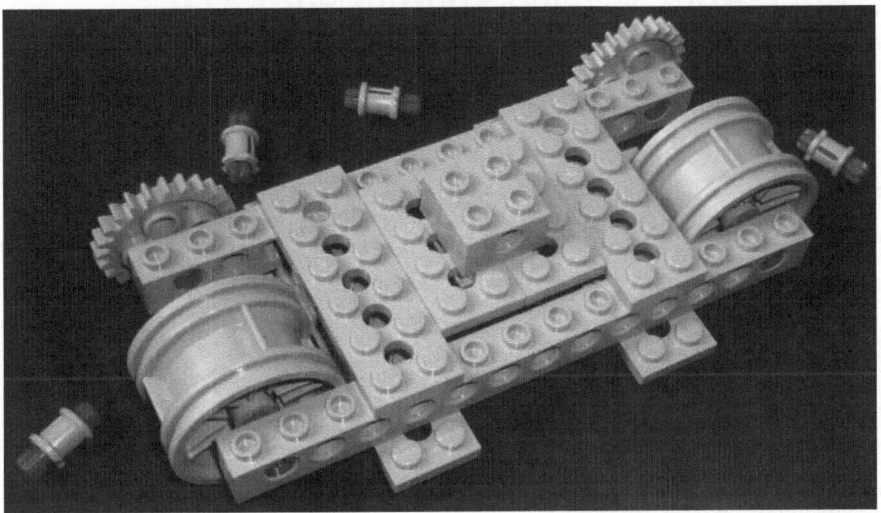

Figure 13-8. Placement of the gel spillage pieces

Scoring the Challenge

The challenge score will be determined using the following system:

- Mission objective 1 completed successfully: 15 points
- Mission objective 2 completed successfully: 20 points
- Mission objective 2 completed successfully: 20 points
- Mission objective 2 completed successfully: 30 points
- Wrist actuator disturbed in objective 1: –25 points
- Index finger breaker switch disturbed in objective 2: –30 points
- Thumb joint assembly disturbed in objective 3: –30 points

Earning Bonus Points

Award bonus points as follows:

- All mission objectives completed successfully with no damage: 30 points

Using the Mars Base Beta Mission Scoring Form

Use the Mars Base Beta Scoring Form to track the completions of the mission objectives and tally the final challenge score.

Note   The Mars Base Beta Scoring Form may be downloaded at `www.marsbasecommand.com` by clicking the Mars Base Beta Scoring button.

Frequently Asked Questions

Following are some frequently asked questions and their answers:

1.   *Do I have to successfully complete all of the mission objectives?*
     No. You may attempt as many of the mission objectives as you wish, but bonus points will only be awarded if all missions are successfully completed without disturbing the three major models.

2.   *What if I successfully remove the damaged battery but do not insert the replacement battery?*
     The replacement battery must be inserted before time expires, or no points will be awarded for the wrist actuator mission.

3.   *Is there a height or weight limit for the mission robot?*
     No. But you are limited in the parts you can use—one NXT robotics kit plus left over parts from one Resource Set (if one is used).

4.   *What happens if my robot disturbs a model? Can I complete the mission?*
     You may complete the mission to offset a portion of the negative points earned.

5.   *What if I encounter a situation that's not covered by the challenge rules?*
     Make your own ruling on the matter that is agreed to by all parties participating in the challenge.

Adding Novice and Expert Rules

As an alternative to the standard mission challenge rules, some novice rules have been created to make the challenge a little easier to successfully complete. These novice rules are as follows:

- The challenge time limit is doubled to 6 minutes.

- The penalties for damaging/disturbing the three primary models is removed; there is no loss of points for this type of action.

As an alternative to the standard mission challenge rules, the following expert rules have been created that will make the challenge a bit more difficult to successfully complete:

- The challenge time limit is cut to 2 minutes.

- Any damage to a model stops the challenge (the doctor calls a halt to the surgical procedure), and points are awarded only for objectives successfully met before the challenge is ended.

Summary

The Mars Base Beta challenge is likely going to be one of the most difficult missions for you. The risks involved with surgery have been simulated here by awarding penalty points for disturbing the sensitive and complex mechanisms involved in the workings of a prosthetic hand.

As with previous challenges, teams and individuals should be given the benefit of the doubt when it comes to interpreting the rules. If a rule is unclear, think about the overall objective of the missions, and make your own decision regarding how you will resolve or enforce that rule.

I've allowed individuals or teams to decide whether to build an autonomous or hand-held robot. Think carefully about this decision, and make the one that will best serve you in terms of learning something new. If you're a novice (individual or team), tackle the hand-held version first, as you won't have to concern yourself with the robot moving and navigating the challenge area. If you've got experience, the autonomous robot is definitely going to push you to learn new design and programming techniques. No one said working for Mars Base Command would be easy, so accept the challenge, and push yourself and your robot to the limits.

CHAPTER 14

Systems Crash

Is It Always Like This Around Here?

Mars Base Alpha, Section D, Commander's Office
August 28, 2062 at 7:14 AM (Greenwich Mean Time)

Lieutenant Kristie Raleigh rubbed her temples and then her eyes. She couldn't decide which was worse—the headache or the lack of sleep. For more than a week, she had helped her team repair the damage done by the micrometeor impacts, assigning staff to duties for which they were not always trained.

"You look like you could use some sleep," said Doctor Kim Homansky, knocking on the open door. "About 24 hours' worth."

Kristie didn't look up but waved her hand to invite Kim into the room.

The doctor took a seat in front of the large desk, watching her friend return to rubbing her eyes and temples. "Seriously, Kristie. If you don't get some sleep, I'm going to have to order you to bed. Don't think I won't do it."

Kristie smiled and nodded. "That'll leave Lieutenant Wozniak in charge."

Kim's eyebrows went up. "That nervous guy from Tyco Station who's always pushing the wrong buttons?"

"Yep," replied Kristie. "Or you could release the commander from sick bay and put him back on active duty."

"He needs rest for his eye and his repaired hand," said Kim. "But you're avoiding the subject. You can't keep operating like this."

"I had three hours of sleep today."

"Yes, I checked the bio logs. That three hours was almost 20 hours ago, so technically, you have slept today. But you're running a severe sleep deficit Kristie. You can't keep pushing yourself like this."

Kristie leaned back in the chair, inhaled deeply and then exhaled slowly. "No choice, Doc. We're finally in the green on 90 percent of our systems, but that remaining 10 percent... ."

"Your people can handle it," interrupted Kim. "Go grab a few hours right now, or I'll have to order you to do so."

"If you're going to push," replied Kristie. "I guess I won't argue."

Kristie leaned forward to stand, but she froze when she heard another knock on the office door. "Lieutenant Raleigh?"

Kim looked over her shoulder. Two of the base personnel were standing in the doorway, Specialist Lisa Camway and Specialist Mason Gerard. Lisa was forcing a smile, but Kim could tell she was nervous. Mason looked like he'd seen a ghost, his eyes wide and face drained of color.

"Yes?"

"Sorry to interrupt you, Doctor Homansky," said Lisa. "But we've got an urgent issue that Lieutenant Raleigh is going to need to hear."

Kim shook her head. "Is it always like this around here?" she asked.

Kristie nodded. "It never ends, even when the base isn't dealing with micrometeors taking out our power and life support systems. Come on in, you two. Pull up some chairs."

Kim scooted her chair to the right and made room for the two specialists. Gerard still hadn't spoken, and Kim was beginning to wonder if he was in shock.

"I'll let Mason explain," said Camway. "He's got a better grasp of the situation."

Kristie looked at Mason. "OK. You have my attention."

Mason swallowed.

"Mr. Gerard. Mason," said Kristie. "Just tell me. Take a deep breath, and let's hear the issue, so we can begin to address it."

Mason nodded. "Uh, yeah. OK, well, you see…"

Lisa shook her head. "Never mind. Lieutenant Raleigh, one of our remote lab satellites took some damage during the storm. We've got two people up there right now, and they're unable to get the emergency escape pod to launch."

"Are they hurt?" asked Kim.

"No ma'am," replied Lisa. "But the satellite has lost its stability and is dropping out of orbit fast."

"Do we have any launches scheduled today?" asked Kristie. "We've got to have an orbiter or two prepped for launch that we can redirect for a rescue mission."

"The earliest launch we can muster will take another fifteen hours. The satellite has about six hours before it reenters the Martian atmosphere."

"My sister is on that satellite," said Mason. "I'm just… I can't…"

Lisa put her hand on Mason's arm.

"It was supposed to be me up there today," replied Mason.

Kristie leaned forward and stared directly into Mason's eyes. "OK, Mason. I know you're stressed, but I need to know more. Do we have communication with the satellite?"

Lisa nodded. "Yes, but Mason's sister, Ginny, and the other tech, Louie, know about the orbital decay. They're a bit stressed right now."

"I would be too," said Kristie. "But let's work the problem. All of you, come with me. Let's do what we have to do to get them safely back to base."

Won't It Just Be a Bumpy Ride?

Mars Base Alpha, Section D, Control Center
August 298, 2062 at 7:22 AM (Greenwich Mean Time)

Lieutenant Raleigh and Doctor Homansky stood side-by-side with Lisa and Mason and stared at the video screen. On it, Ginny Gerard and Louie Pitchersky floated in zero gravity, surrounded by various types of equipment and research bays.

"Have you tried rebooting the satellite's main computer system?" asked Kristie.

"We did," replied Louie. "No luck. The escape pod will not power up. We can release the docking clamp manually, but unless the life pod's system boots up, we won't have control of its angle of descent."

"What do you think the problem is? Any ideas?" asked Kristie.

"I stuck a scope under the dashboard," said Ginny. "I'm not a computer expert, but I think I spotted some burned areas. The life support system is keeping the air clean, but I'm pretty sure I can smell a bit of burned ozone."

Kristie nodded. "Alright, I'm going to get a systems expert up here right now." She nodded at one of her staff members, who turned and began typing on a keyboard. "Just one minute."

Kim spoke up. "How are the two of you doing? I know stress levels are high, but are either of you experiencing any dizziness? Are you having difficulty breathing?"

Ginny turned to Louie who shook his head. "No, ma'am. Nothing like that. We'd just both like to be on solid ground, if you know what I mean?"

"Just keep a calm head," replied Kim. "Lieutenant Raleigh is doing what she can, but I need the two of you to try to calm yourselves and stay focused."

Kristie turned her attention back to the screen. "I've got someone on the way right now. Before he gets here, let's do a quick inventory, OK?"

Both specialists nodded on the screen.

"What kinds of equipment do you have with you? We're going to need to look for possible replacement parts, so take a look around and see what you can come up with. Some of that equipment behind you looks promising."

Ginny shook her head. "There is some computational equipment up here, but the problem is that the emergency escape pod is all black-box interfaces. I can't get into them."

Kristie frowned. "There should be a toolbox mounted to the right of the primary maintenance panel."

"It's there," said Ginny. "But it's all macro tools—screwdrivers, wrenches, that kind of thing. I can remove the dashboard, but I don't have any tools to crack open the black boxes to inspect them."

"There are no backup controls?" asked Kim. "Someone had to have considered spare parts for the escape pod, right?"

Kristie nodded. "The redundancy is built in. Each black box is a separate control unit. There should be three of them if I'm not mistaken."

"Yes, there are three," said Ginny. "I can get to them easily enough, but I have no way of knowing if one or all of them are damaged."

Mason turned to Kristie "Can't we just stabilize the satellite's orbit until a rescue attempt can be made?"

Kristie turned to a technician who was monitoring the satellite's orbit. "Sergeant Lamm, what's the status on the satellite?"

The man shook his head. "The satellite's orbit is beyond repair. Firing the thrusters won't have any affect. I'm sorry. The estimate for reentry is 5 hours and 22 minutes."

Kristie turned back to Mason. "OK, so fixing the satellite is not an option. We've got to get the escape pod powered up."

"But if the systems have totally crashed and they don't have any replacement black boxes, can't they still launch the escape pod?" asked Lisa. "Won't it just be a bumpy ride to a random landing spot?"

"I'd suggest we not test that theory," said a winded voice behind the group.

Kristie turned and smiled at the out-of-breath figure standing a few feet away. "Sorry to wake you up, Henry."

The man's hair was standing up, and he still wore a pair of pajamas, which were sticking out from the bathrobe he'd tied at the waist. "I won't be able to sleep again until we get these folks home," said Henry Cho, Mars Base Alpha's lead computer systems expert.

A Very, Very Tiny Pair of Pliers

Mars Base Alpha, Section D, Control Center
August 28, 2062 at 8:18 AM (Greenwich Mean Time)

It had taken Henry an hour to walk Ginny and Louis through a series of reboots and system checks, but the escape pod still would not power up. The dashboard on the pod had been removed, and various wires were pulled and tested with a multimeter to ensure that that voltage was not the issue.

"What types of experiments are running on this thing?" asked Henry as he examined the schematics of the escape pod.

"Standard stuff, this time," replied Kristie. "Some gravity measurement experiments, a couple of chemistry boxes testing various reactions under zero g. Plus whatever experiments Louie and Ginny took up there but haven't logged into the system yet—some microbot tests away from their lab. But no large-scale experiments requiring anything more than the satellite's on-board computer system to maintain."

"There's no way to transfer the satellite's computer to the escape pod?" asked Mason.

"Impossible," replied Henry. "It's large and would take too long to properly disassemble."

"Wait," said Kristie. "They have their data pads. Can't we interface those?"

Henry shook his head. "You're going to scream when I tell you this. Different operating systems."

"I'll scream later," replied Kristie. "What are our options at this point?"

Henry pointed at a pad of paper where'd he been scribbling notes and scratching out ideas that didn't work. "Without replacements, it all comes down to finding out what damage has been done in those black boxes. Two of the boxes are obviously beyond repair. The one on the far left won't give me any feedback. It's toast. The middle one is responding, but it appears that some sort of electrical damage has been done to the CPU. With the right tools, I could crack that box and replace the CPU, but . . . "

"No tools up here, and you're down there," replied Louie.

"Yes. Sorry," said Henry.

"What about the third black box?" asked Lisa.

"The diagnostic port is sending back three specific error codes. The first is telling me that the interface port inside the box has become disconnected or burned out. An easy fix if they had the right tool to open the black box. The second code indicates that the primary memory module is damaged, but there's a simple toggle switch inside the box that would flip it over to the backup module."

"And the third code?" asked Mason.

"That's the big one," said Henry. "The heat sink for the processor has become detached. The tiniest blob of solder or any kind of conductive material would pull the heat away from the CPU and direct it to the cube's exterior. Even if we fix the interface port and switch the box over to backup memory, that CPU will burn out the moment we power it up if the heat isn't dissipated immediately. The CPU is hard-coded with the calculations needed for the escape pod to safely return to Mars."

"This is crazy," said Mason. "All they need is to open the black box, but they don't have the right tool to do it."

"What about the tools they do have?" asked Kim. "I've been forced to use some surgical tools in ways they weren't designed. How about cracking the case with a pair of pliers?"

Henry sighed. "These black boxes are tiny—only about half a centimeter in width and height, and less than a quarter of a centimeter in depth. The human hand just doesn't have the fine control to crack the box without damaging everything inside."

"What if I use a very, very tiny pair of pliers?" asked Louie with a grin as he tried to lighten the mood.

"You'd want a very, very tiny pair of hands to do it," responded Henry. "Sorry, Louie, but we've got to look elsewhere for a solution"

"Hold it," said Mason. "I think you've just given me an idea."

Everyone turned to face the specialist who now had a large grin on his face.

Think Small

Remote Lab Satellite 23
August 28, 2062 at 10:17 AM (Greenwich Mean Time)

"I think it'll work." Ginny grinned at the screen. "It's a good thing we chose not to program the microbot down there, Mason."

"Why is that?" asked Henry.

"Our microbots use a special type of memory. Once the code is uploaded, it can't be changed. It's a security option we implemented to keep them from being reprogrammed."

Louie floated over and bumped against Ginny as he tried to get in the camera's field of vision. "How will we know if the little fellow is successful?"

Henry frowned. "If the microbot can repair the interface port first, we'll be able to monitor the memory module and CPU heat sink repair before flipping the switch."

Mason stepped into view. "Ginny, I've uploaded the program to your data pad. It'll single out one microbot and apply the special programming."

"Wouldn't it be better to send in more than one microbot? Program one group to tackle the interface port and other groups to fix the other issues?" asked Lisa.

"These aren't nanobots; they'll get in the way of each other," replied Mason. "Think small, but not that small."

Ginny had turned her back to the screen and touched her data pad to a small white box floating to her right. She tapped a few buttons on the touch screen, and it beeped. "Looks like the program loaded. Microbot 142 has responded and is waiting to go."

Henry examined a drawing of the microbot. "It will need to enter the black box near the interface port. That's the only space large enough to allow the microbot inside the cube."

"Got it," said Ginny.

Kristie consulted with the technician monitoring the satellite and turned back to the screen. "You need to get moving on this. You've got less than three hours. If the microbot can't fix the black box, we'll need to find another solution."

"Alright, I'm going to instruct 142 to exit containment and signal us when it's ready to enter the black box," said Ginny.

Louie pulled back some wires and shifted the dashboard pieces to give Ginny more room to operate in the small escape pod. He was floating upside down, his head to Ginny's head, and staring at the small white box she carried.

"Here we go," said Ginny as she touched the white box to the small black cube.

Your Turn

Mars Base Beta
August 28, 2062

Were Ginny and Louie able to successfully power up the escape pod in time? Their survival depends on you. Continue reading Chapters 15 and 16 to build the models that will be used in this last challenge. After the models are completed, read Chapter 17 for the rules and challenge setup instructions. Last, you'll need to begin brainstorming a solution to build your own microbot that will enter the black box and perform the repairs that will allow the escape pod to power up and safely return the technicians.

Good luck!

CHAPTER 15

■ ■ ■

Setting Up the Black Box Challenge

To successfully complete this fourth Mars Base Alpha Challenge, you will be required to construct a microbot that will enter the damaged black box and make repairs to enable the escape pod to carry its two passengers safely back to the planet's surface. The black box has three issues that must be resolved:

- Repair the interface port.
- Switch from memory module 1 to memory module 2.
- Reconnect the heat sink to the CPU to prevent chip burnout.

To simulate the black box and the three actions that must be performed to repair it, you will need to assemble the models found in this chapter and in Chapter 16. Figure 15-1 shows the finished collection of models that you will be building.

Figure 15-1. The models that will be featured in the Black Box Challenge

In addition to repairing the black box, the microbot must remove itself from the black box when it's finished without disturbing any of the components. If the microbot remains inside the black box, there is a possibility it may damage one or more components during launch.

Tackling the Black Box Challenge

Before you begin brainstorming how you'll attempt to complete this challenge, you'll need to assemble all the models from this chapter and Chapter 16. Having the models in front of you will allow you to measure the components, inspect those that need to be touched and moved by your robot, and place them in the proper locations with respect to one another to help you with determining the proper size of your robot for navigating the challenge area. Once you've placed the components in the challenge area, you'll want to examine them further to determine the proper actions that must be made to complete this challenge.

You can find more details about the individual models and the tasks involving them in Chapter 17, but one of your goals right now should be to start brainstorming the size and shape of your final robot. The black box is small, and space for movement is limited. Your robot's success (or failure) will be based on how well it moves inside the box and if it is capable of repairing the damage without disturbing any key components inside the black box.

The robot in the story is referred to as a microbot for a reason—the smaller the robot, the easier it will be to navigate around the black box without disturbing other components. Your robot can touch some of these components, but too much movement will result in penalty points.

Because your robot must stay within the boundaries of the black box, penalty points will also be awarded if any portion of your robot touches the boundary that defines the black box's walls. Your robot may also only enter and leave at one location on the black box, so keep that in mind when you decide in what order you intend your robot to execute its repairs.

Although three tasks must be completed for this challenge to be considered successful, a couple of subtasks can be accomplished for bonus points; these will be explained in Chapter 17.

This challenge is tough! Your robot design must be small in form but have the ability to perform some unique actions without damaging the sensitive electronics inside the black box. Do not limit yourself to one robot design; instead, try to improve your design during the testing phase to reduce its size, improve its movement accuracy, and increase its speed by finding the best order to conduct repairs.

Examining the Challenge Area

The microbot must be truly autonomous in this challenge. Once it enters the boundaries of the black box, you will not be able to touch the robot until it leaves the black box. It is possible for you to have the microbot leave the black box so you can touch it, swap out attachments, upload different programs, and other tasks. However, you'll find that sending in your robot once to conduct all repairs will provide the most points if it is successful in completing all three primary objectives.

Read Chapter 17, and examine the challenge area carefully. Make note of the areas where your robot may enter and leave, as well as the tight areas (mainly the corners) where your robot may cross the boundary of the black box and receive penalty points.

The challenge area not only has the three models that your robot must interact with, but it also contains some components unrelated to the challenge that will be referred to as "black box minor components." While your robot isn't required to do any work on these other components, these minor components do offer a further challenge because you'll want your robot to avoid disturbing them. Subtle movement of these components is allowed, but major disturbances will result in penalty points. You can find more details about the black box minor components in Chapter 17 as well.

Building the Interface Port and Memory Switch

The first half of the building instructions for the black box challenge are provided in the remainder of this chapter. Each image shows the pieces you will need to locate in the LEGO Education Resource Set and their quantities. You will also see where these pieces are to be placed.

Chapter 16 will continue with the building instructions for the remaining models, the heat sink and the minor black box components. You'll take your assembled models and proceed to Chapter 17 to set up and run the challenge.

Note If you are uncertain about the placement of a piece in a figure, jump ahead to the next image or even go back to a previous image to make certain you've built the model correctly so far. Examine the figures carefully, and you should be able to correctly identify the pieces and their final locations on the models.

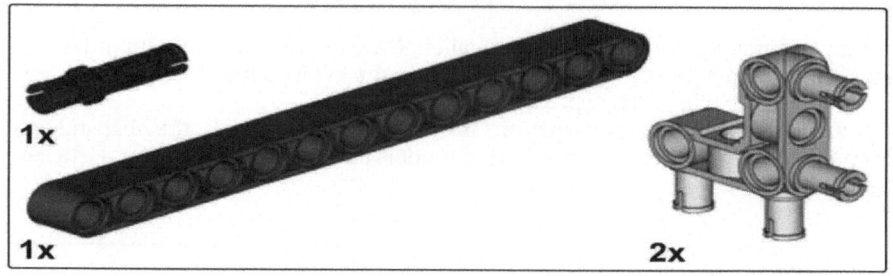

1

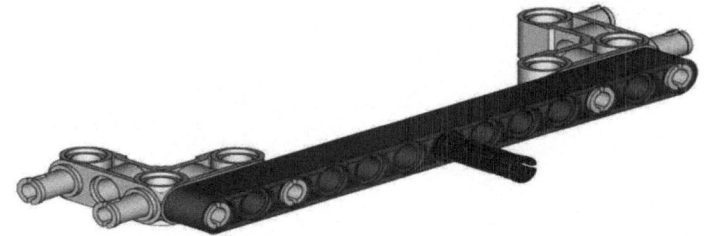

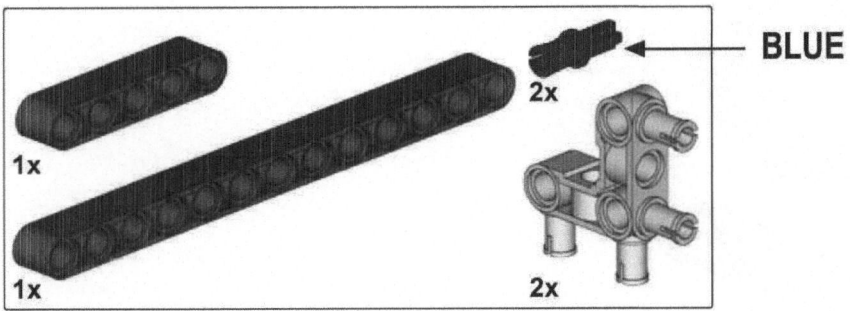

BLUE

2

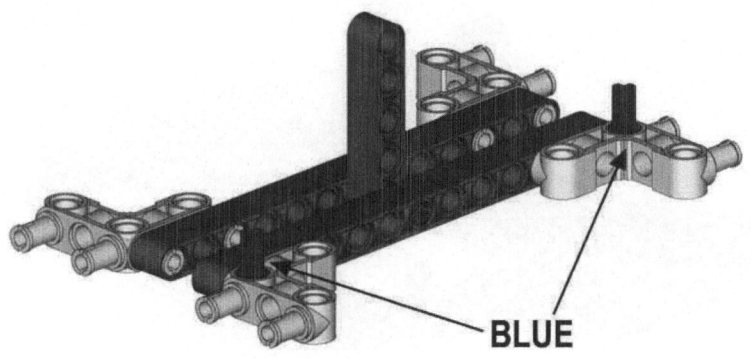

BLUE

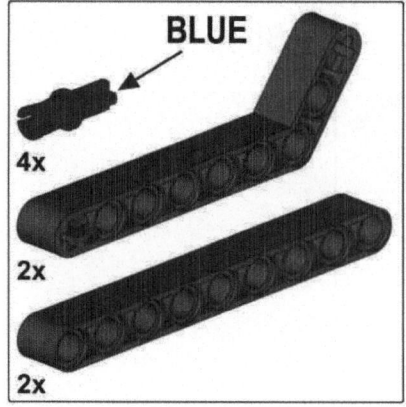

3

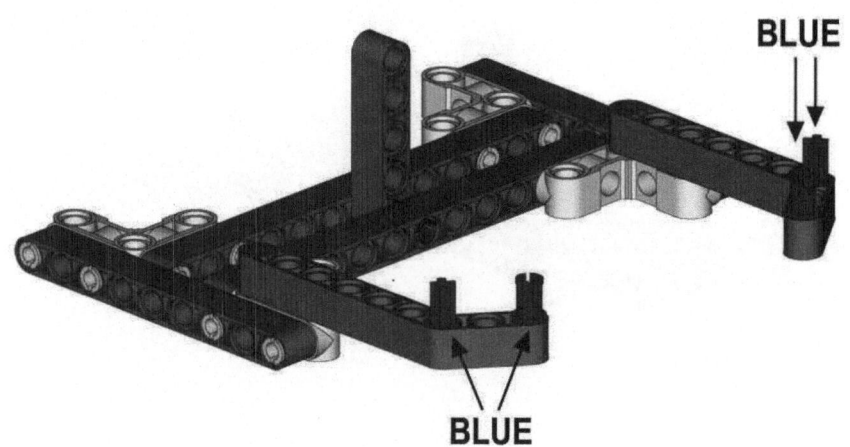

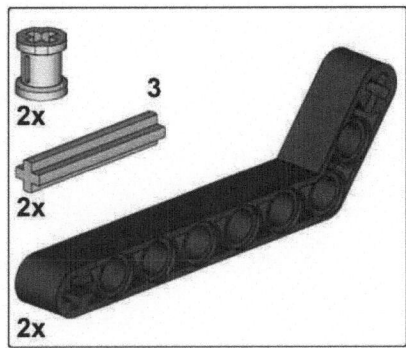

4

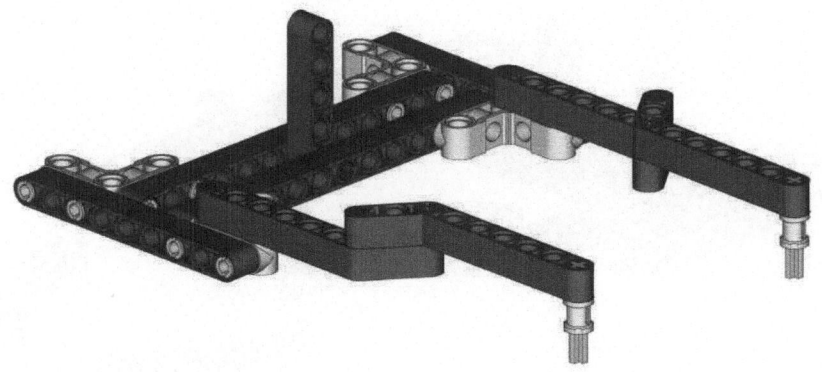

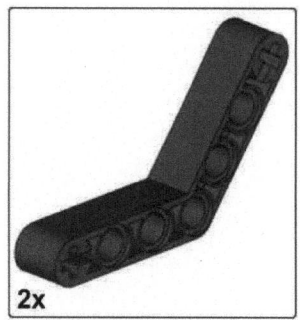

5

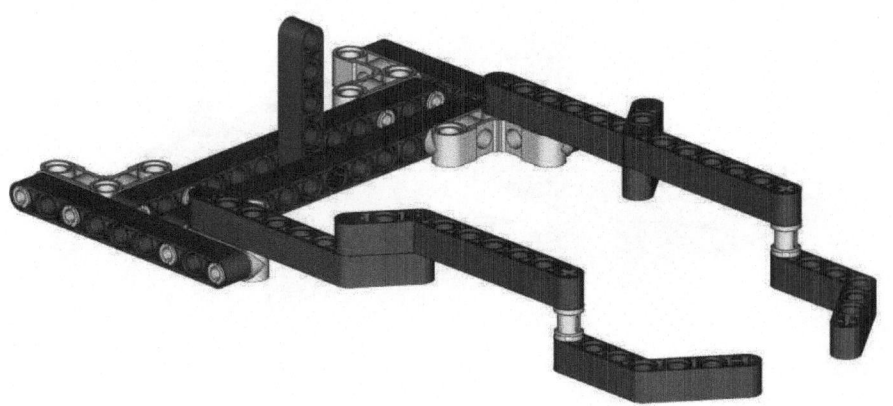

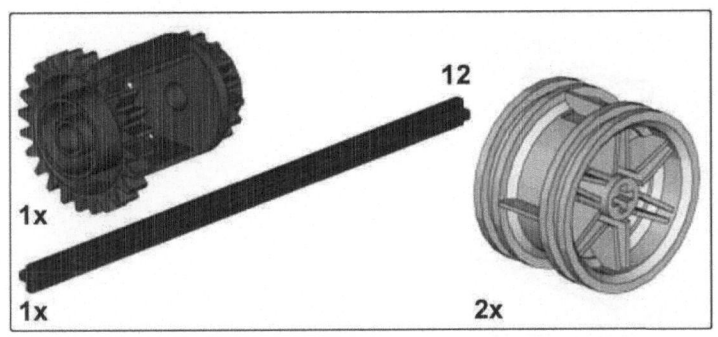

12

1x

1x

2x

6

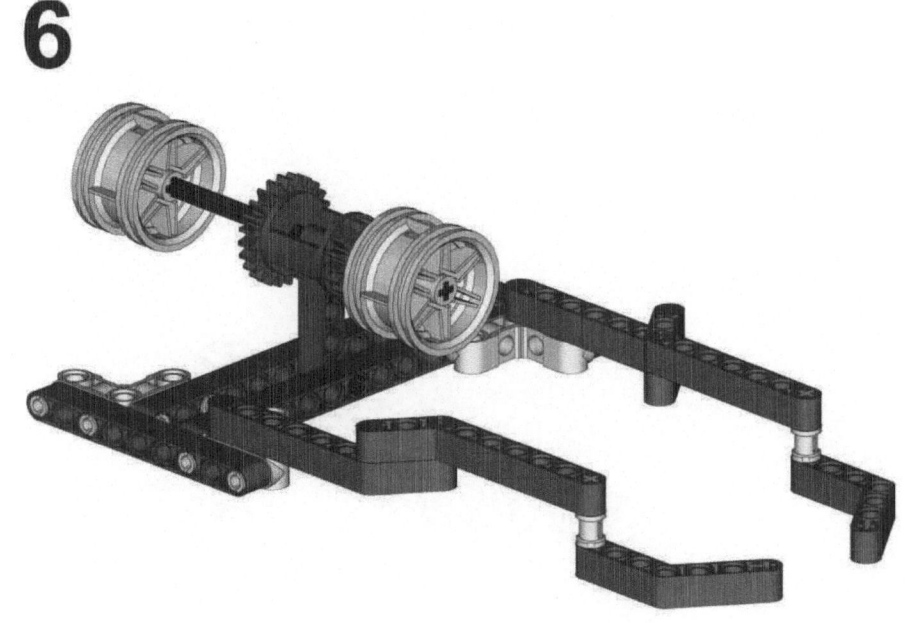

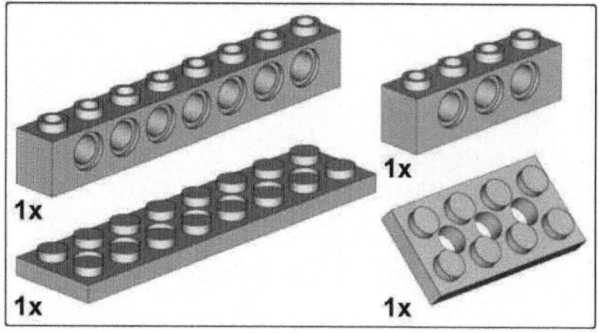

7

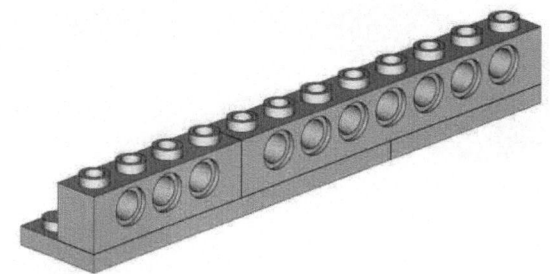

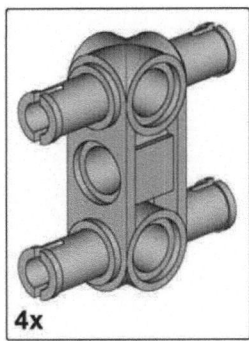

8

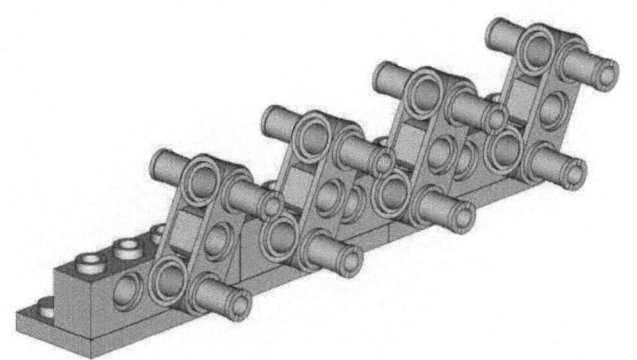

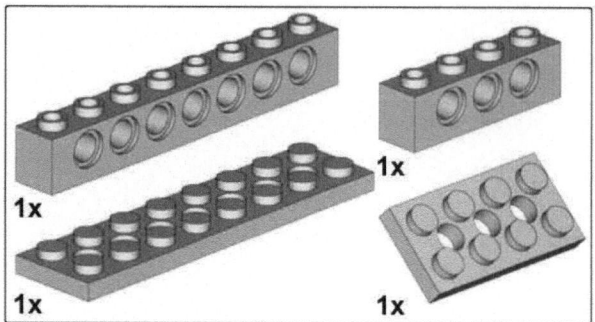

9

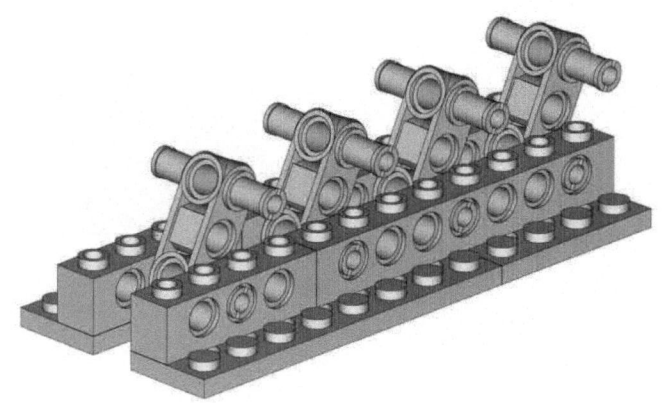

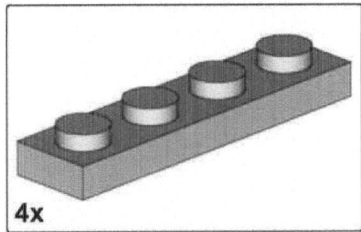

10

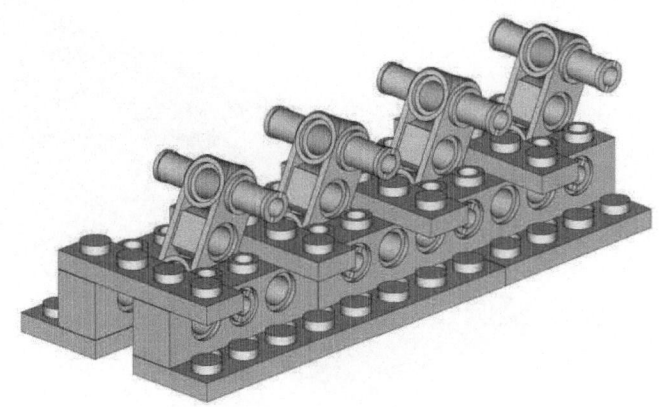

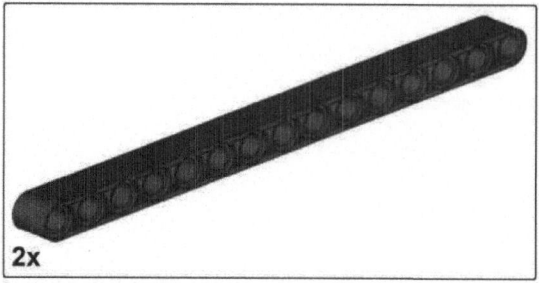

11

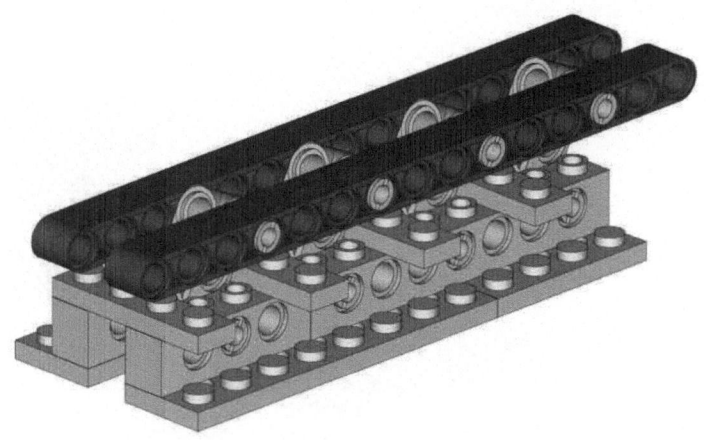

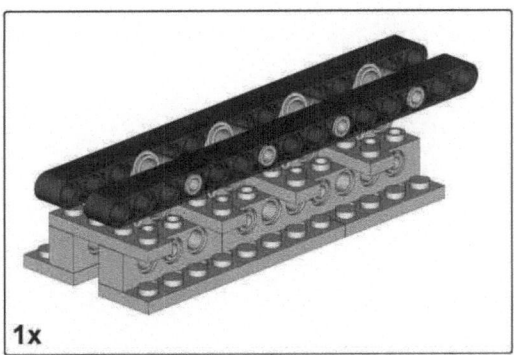

12

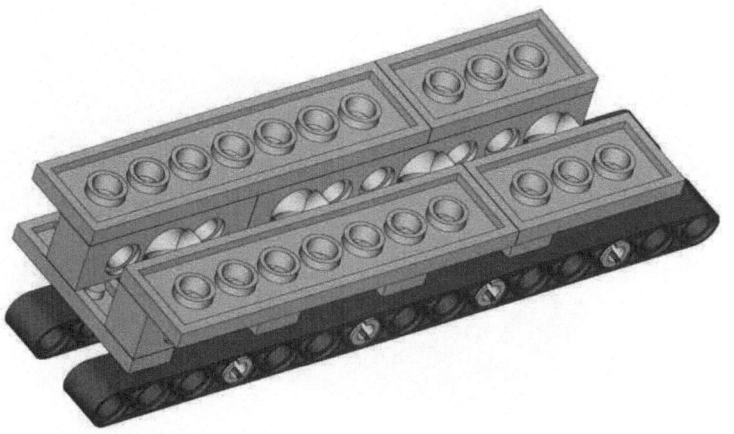

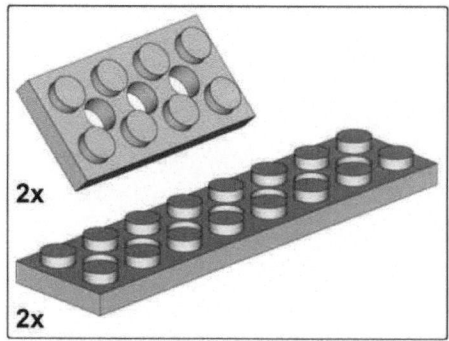

13

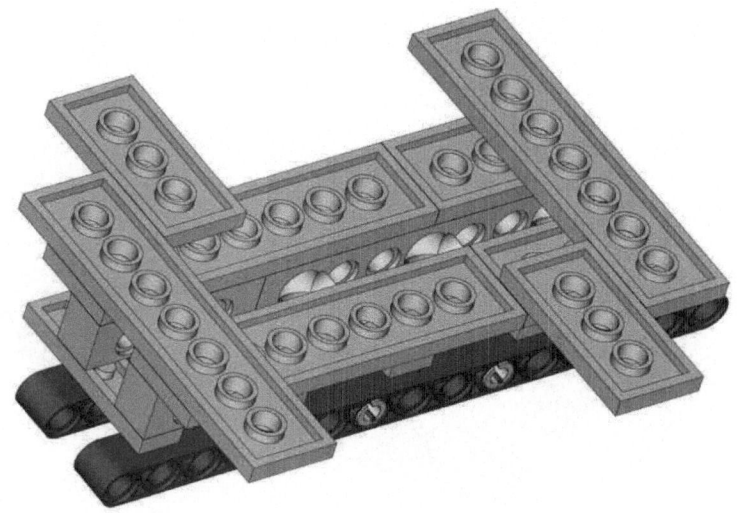

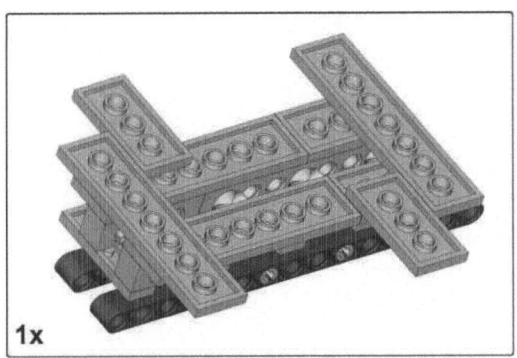

14

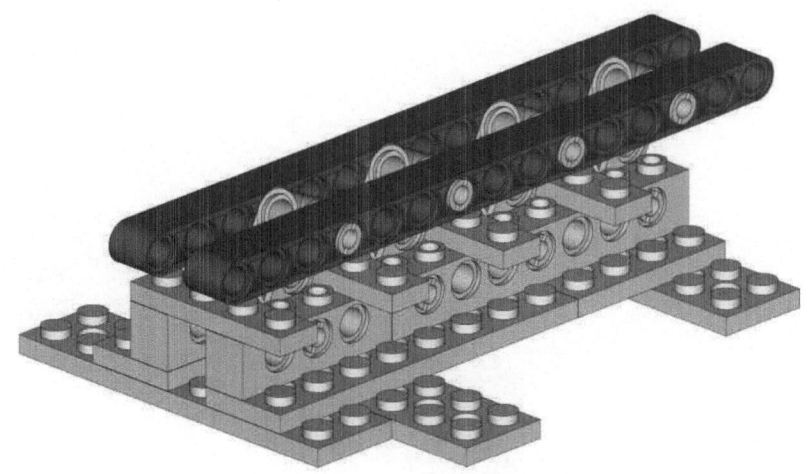

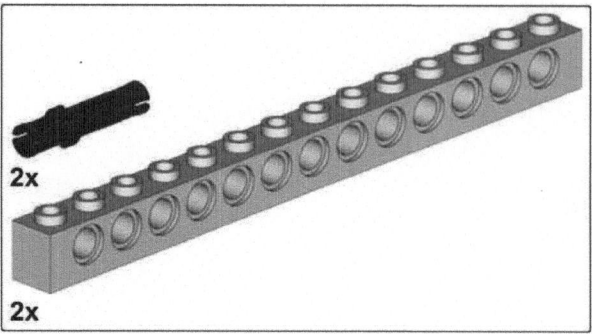

15

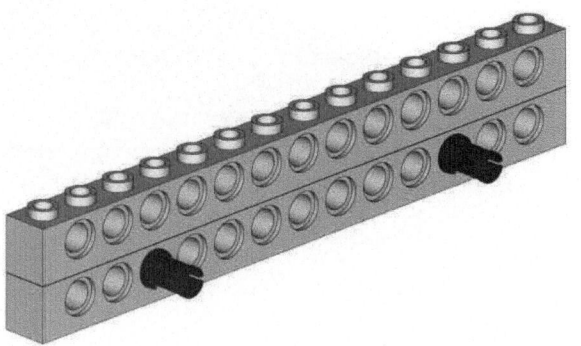

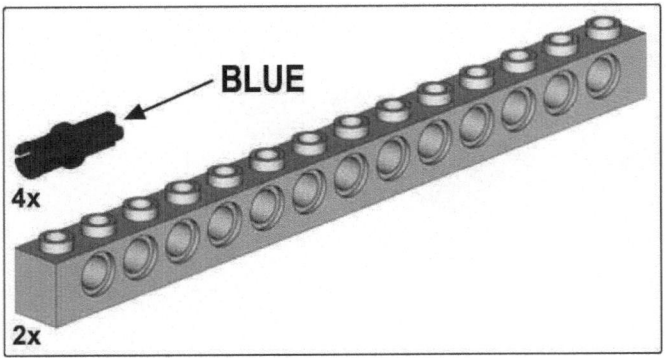

16

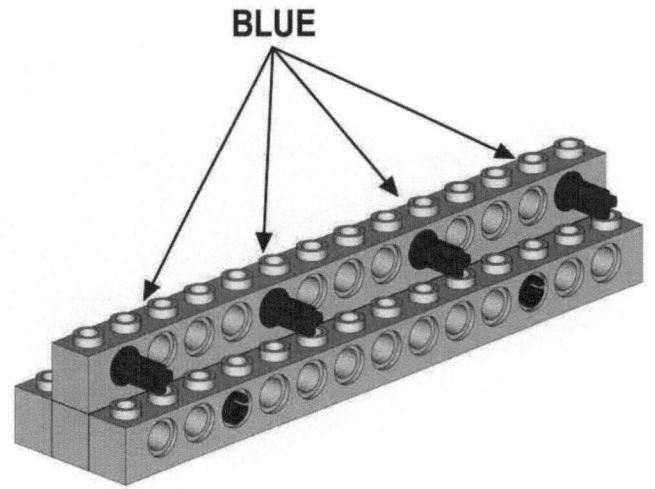

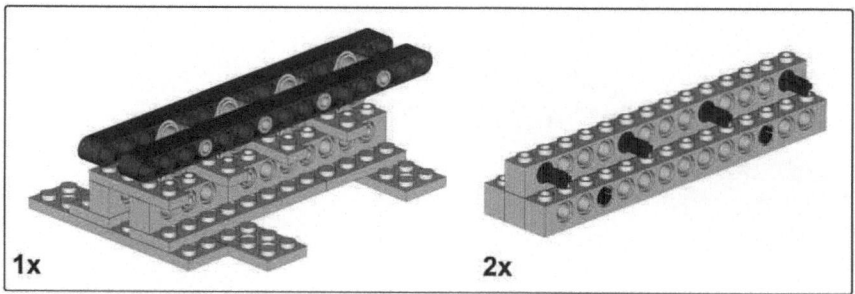

17

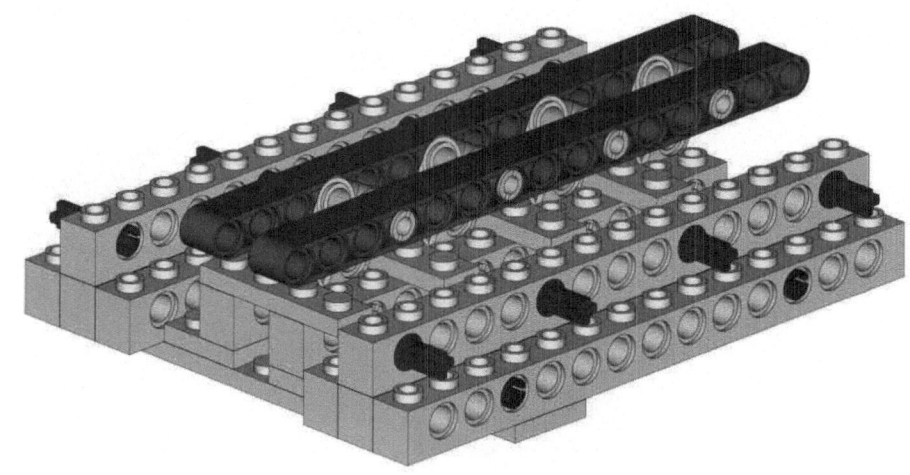

Summary

You're halfway done with building the models for the fourth challenge.

CHAPTER 16

■ ■ ■

Finishing the Black Box Challenge Setup

You'll find the building instructions for the heat sink model as well as the black box minor components used in the fourth challenge in this chapter. Each image shows the pieces you will need to locate in the LEGO Education Resource Set and their quantities. You will also see where these pieces are to be placed.

After you finish building the models, you'll need to read Chapter 17 to learn the rules of the challenge and how to set up the challenge area.

Note If you are uncertain about the placement of a piece in a figure, jump ahead to the next image or even go back to a previous image to make certain you've built the model correctly so far. Examine the figures carefully, and you should be able to correctly identify the pieces and their final locations on the models.

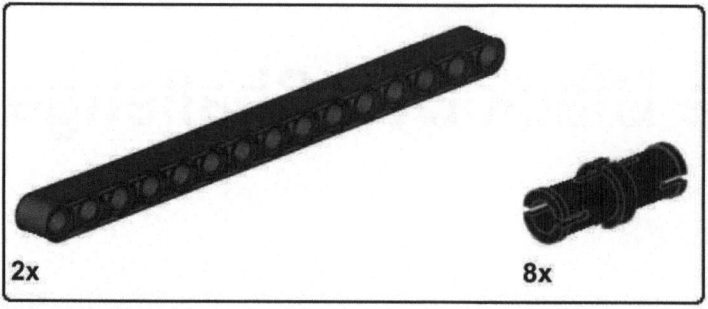

1

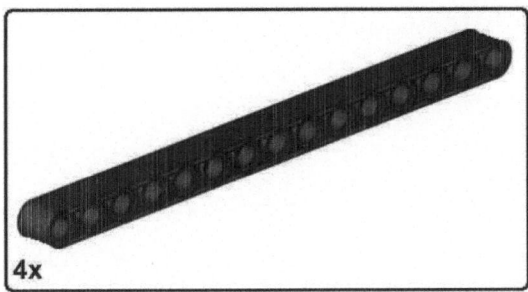

4x

2

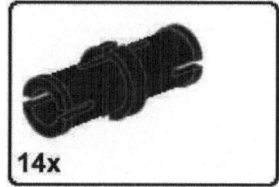

3

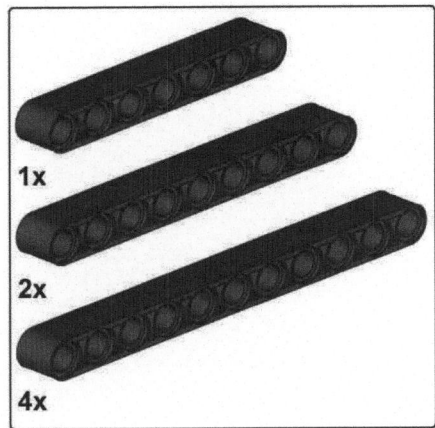

4

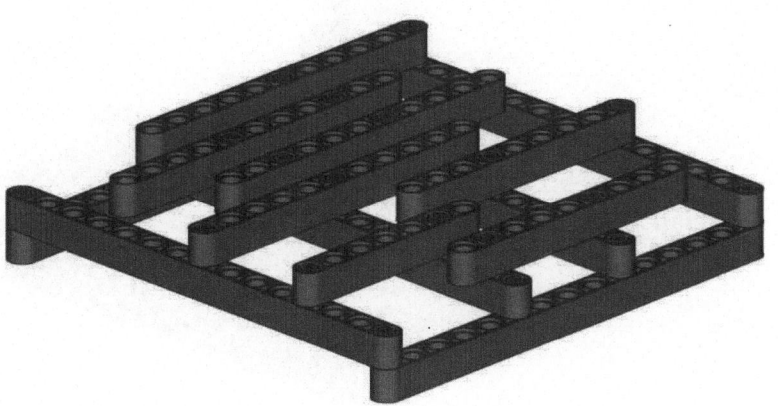

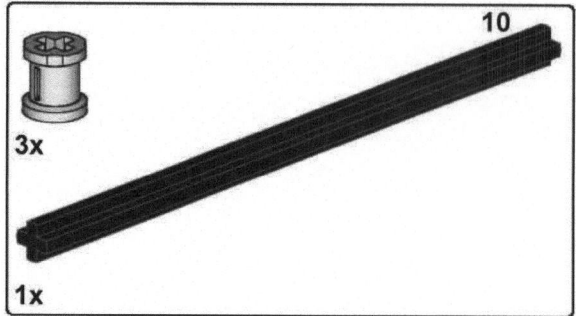

5

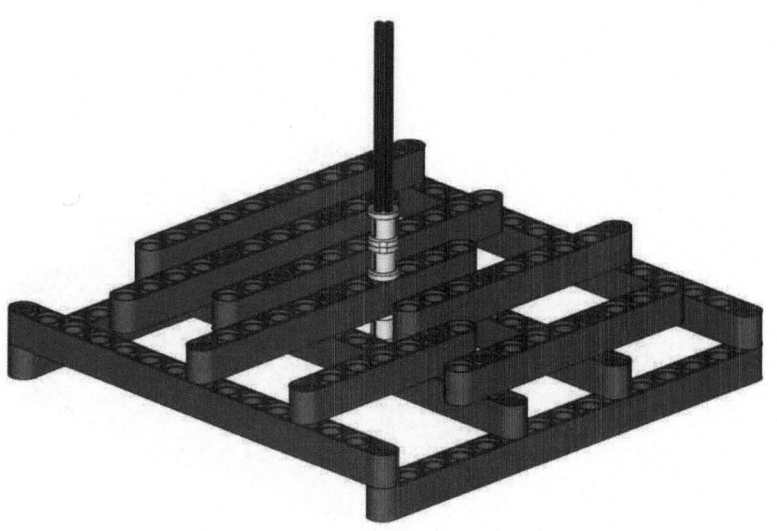

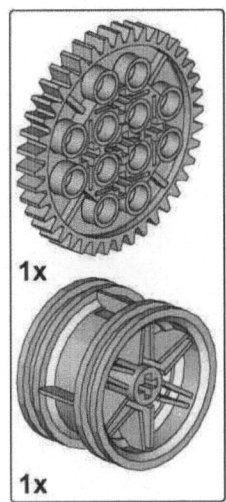

6

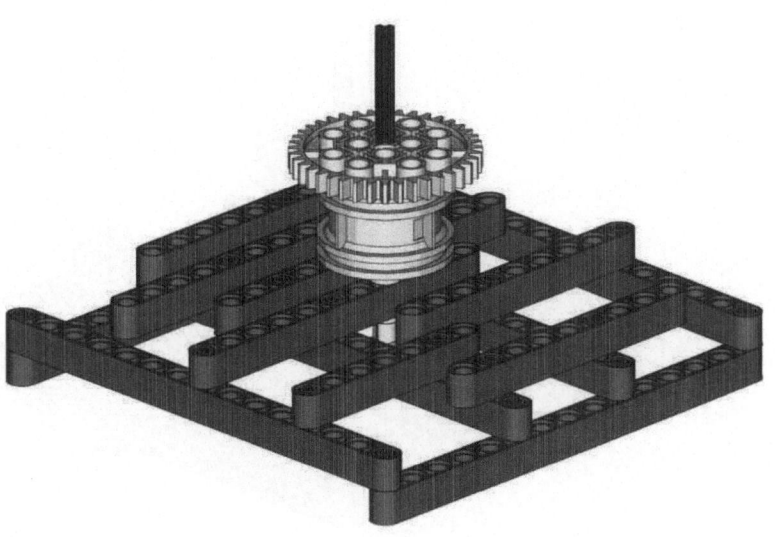

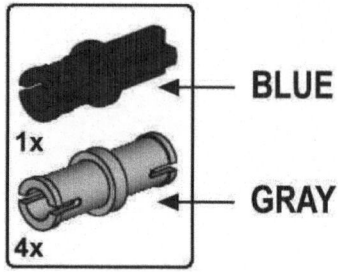

BLUE

1x

GRAY

4x

7

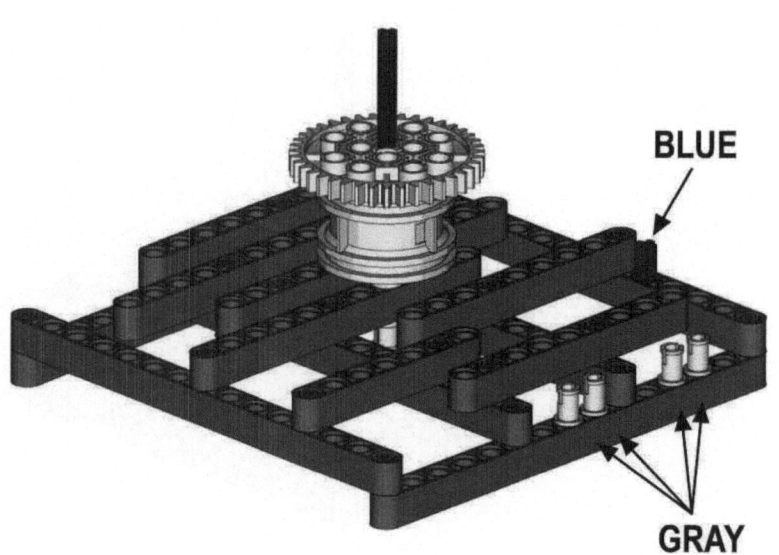

BLUE

GRAY

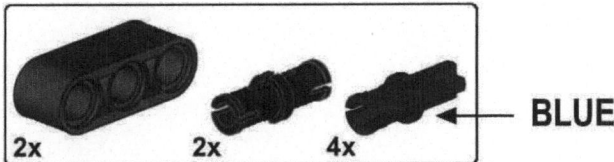

BLUE

8

BLUE

BLUE

9

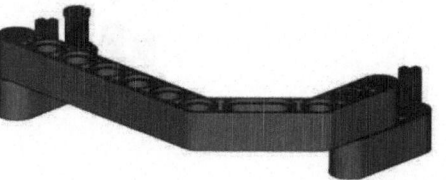

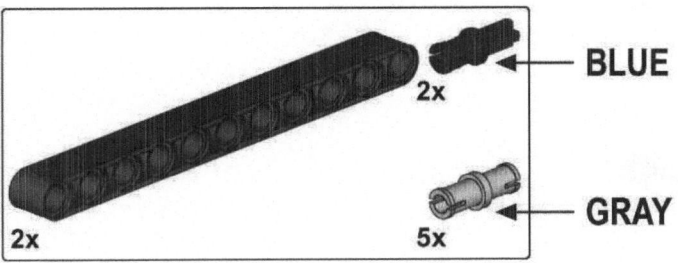

10

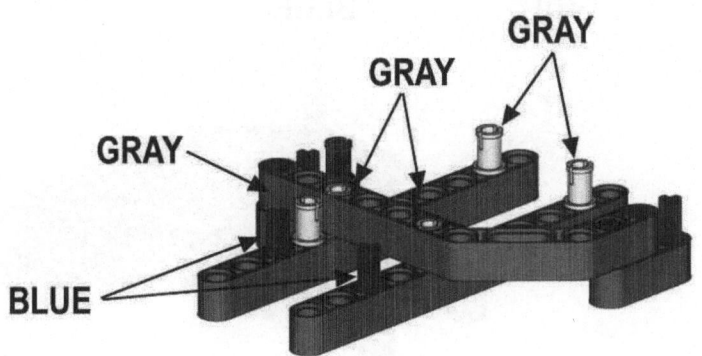

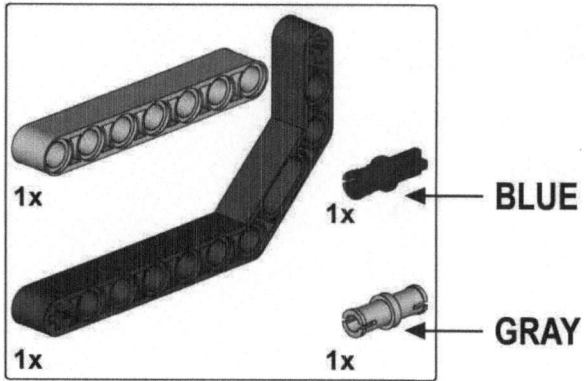

11

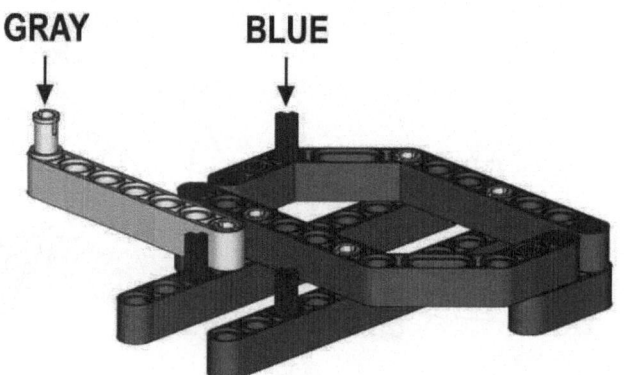

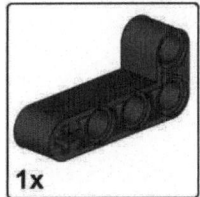

12

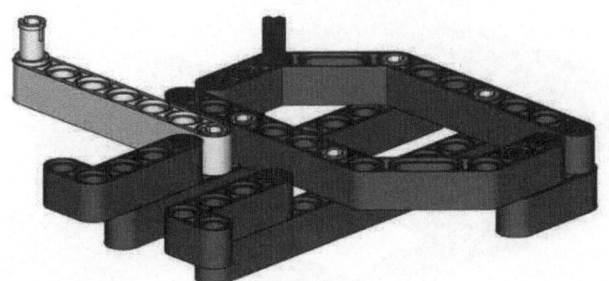

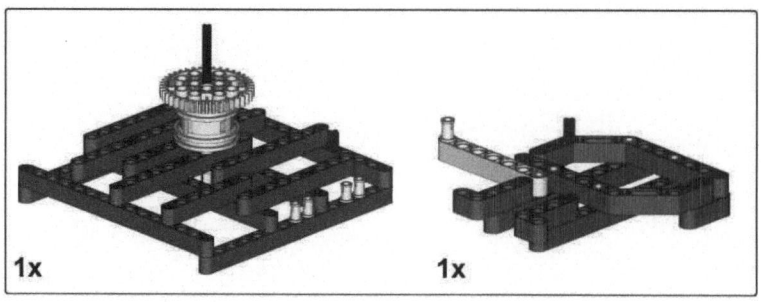

13

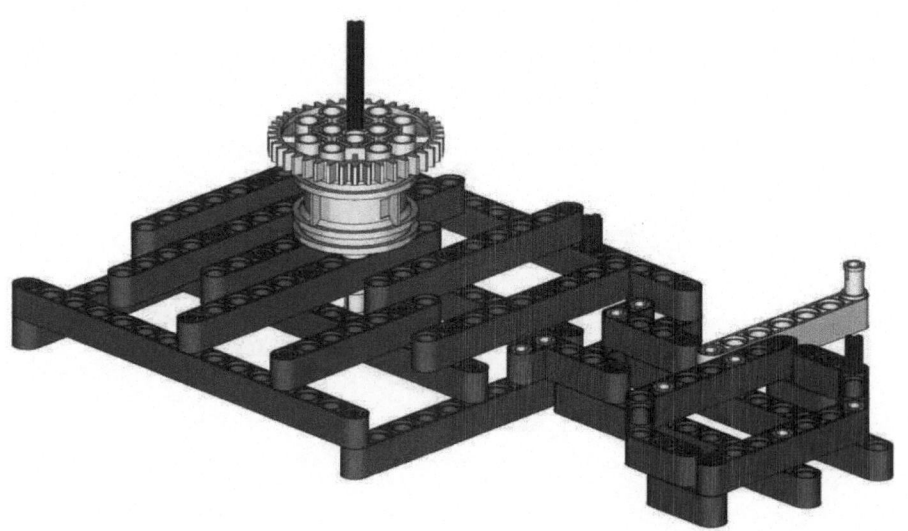

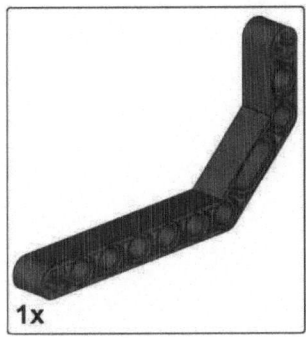

14

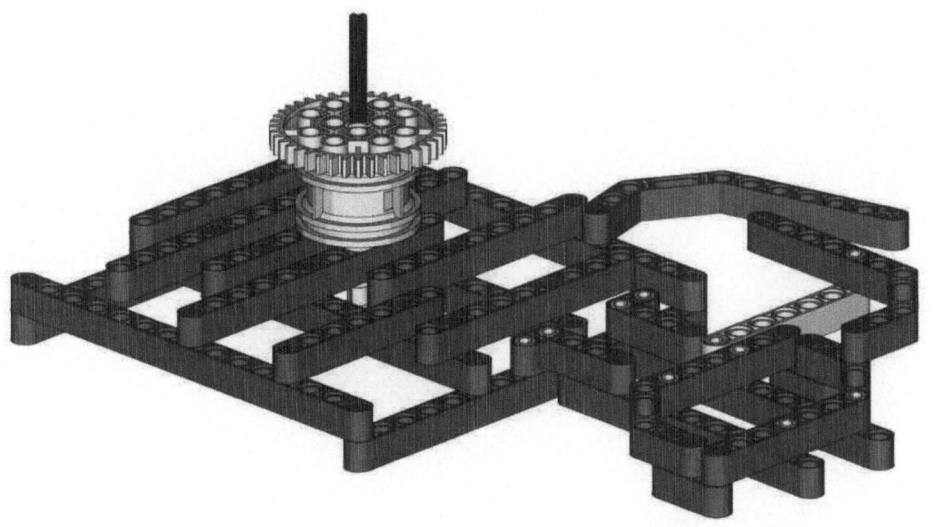

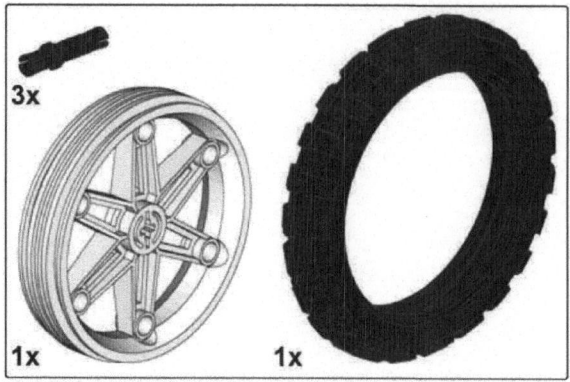

15

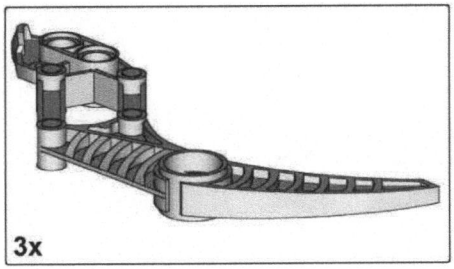

16

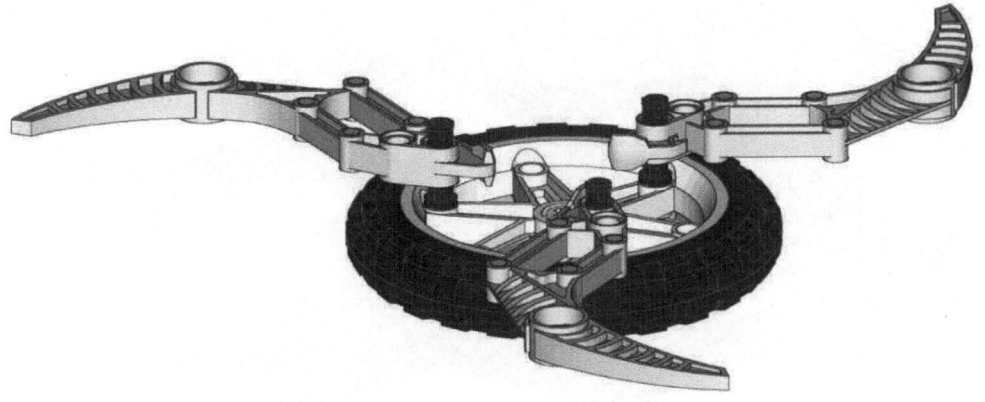

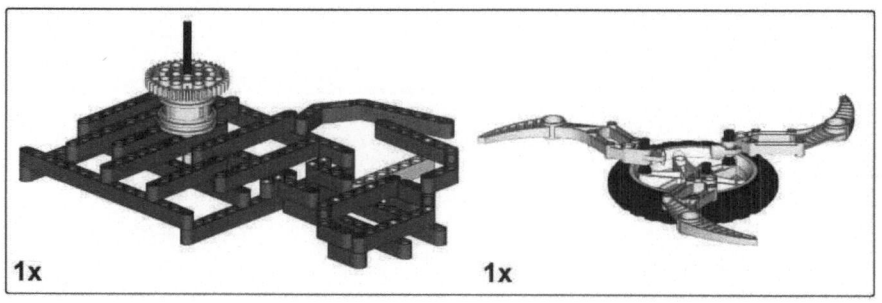

17

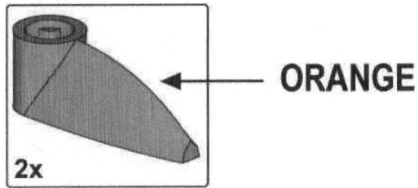

ORANGE

18

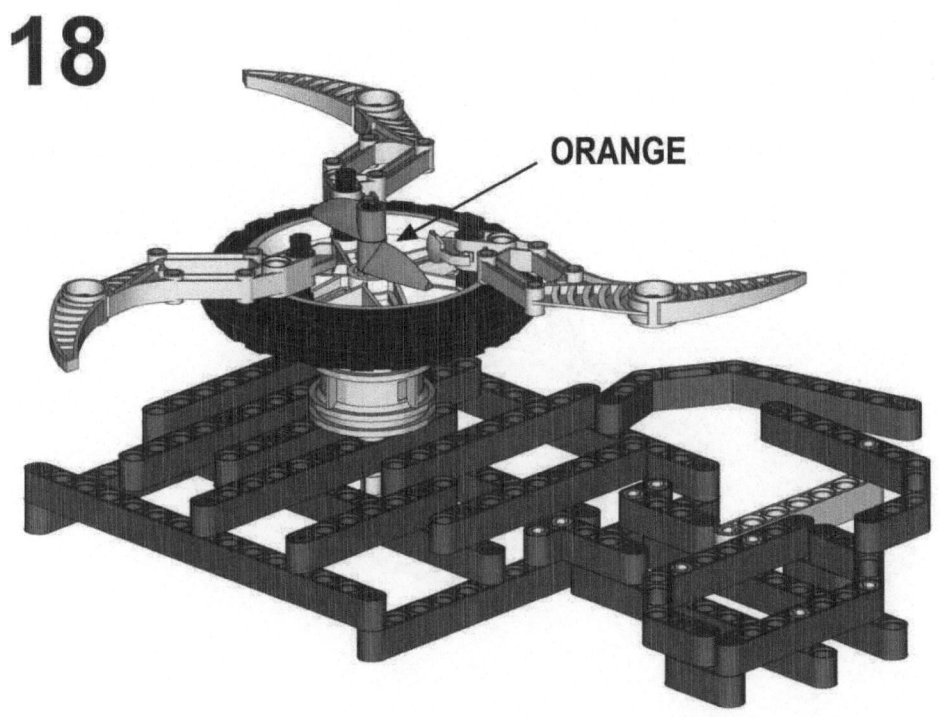

ORANGE

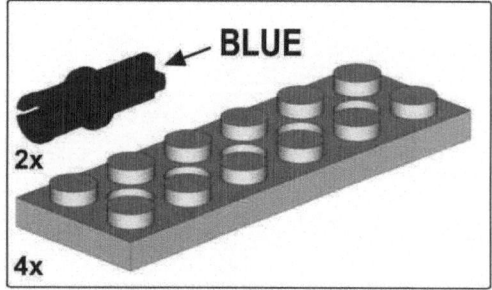

19

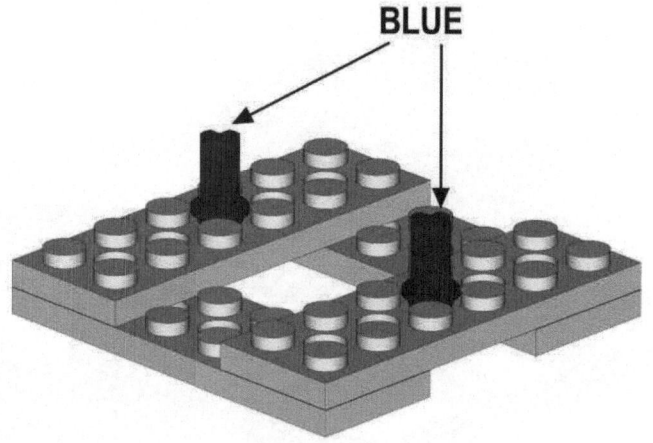

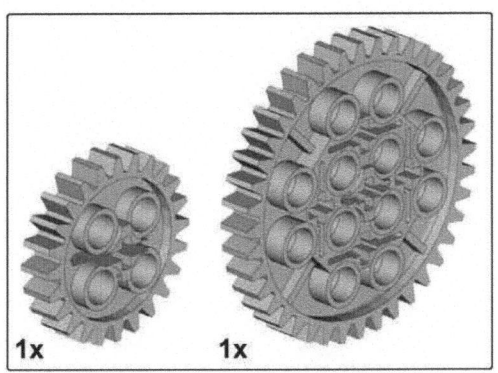

20

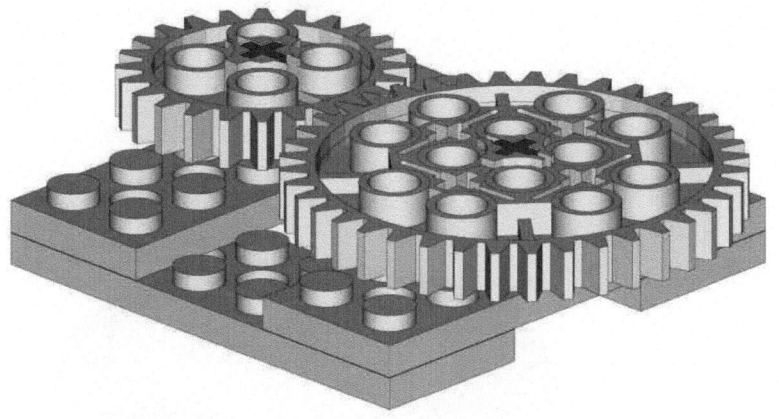

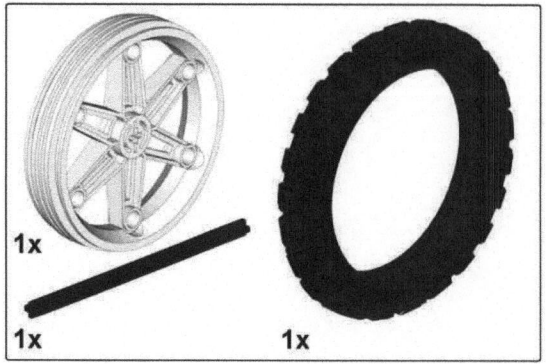

21

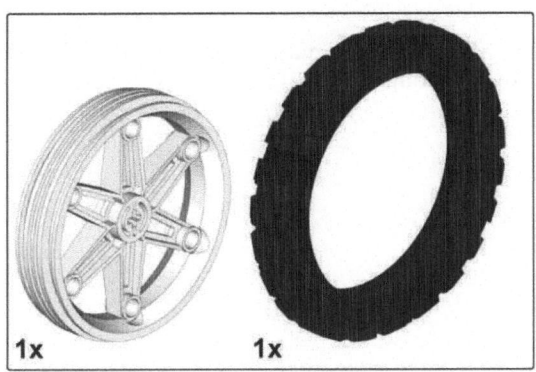

1x   1x

22

Summary

Now it's time for you to complete the final challenge at Mars Base Command. Continue on with Chapter 17 to learn how to setup the models to simulate the final challenge.

CHAPTER 17

■ ■ ■

Synopsis and Rules of the Black Box Challenge

The Mars Base Alpha: Systems Crash challenge will push your robot building and programming skills to their limits as you attempt to build a microbot that will enter the black box and make repairs to the electronics components inside.

Teams or individuals will build and program a LEGO MINDSTORMS NXT robot that must perform repairs in a small space and while avoiding other critical components. Three primary objectives must be completed to successfully repair the black box and award full points, and two secondary objectives will earn bonus points for players who are able to add a bit more functionality to their robots.

To successfully complete the Mars Base Alpha Systems Crash challenge, your robot will need to perform the following actions:

- Repair Interface Port 2 by ensuring that the plug port is pushed so that it extends beyond the black box's boundary. Doing so will offer the black box the ability to communicate again with the escape pod's other computer systems.

- Switch the black box from using damaged Memory Module 1 to using the backup Memory Module 2.

- Reconnect the heat sink so that the heat from the CPU is properly dissipated to the walls of the black box.

- Avoid two other critical components inside the black box: the current-leveling capacitor and its kinetic charging gears.

Points will be awarded for each repair that is successfully made, and bonus points awarded for complete repair of the black box. Points will be lost should the robot move beyond the boundary of the box while making repairs or if the robot comes into contact with the black box secondary components, the Capacitor or the kinetic charging gears (there are exceptions to these rules, and you can read more details later in this chapter).

Setting Up the Challenge Area

The challenge area (CA) that defines the boundaries for your robot must adhere to the restrictions explained in this section. Use Figure 17-1 as a guide for proper placement of the models, or download and print the Challenge Area Mat PDF file.

Note   You may download the full-color PDF and black and white PDF mat files for use in this challenge by visiting www.marsbasecommand.com and clicking the Mars Base Beta Mat button. The files may be taken to a printer and printed as a mat (color or black and white) to be used for the challenge area. Permission to download and use the file for private use is granted. The file may not be sold, and printed mats may not be sold. *The mat is not required to run the challenge.*

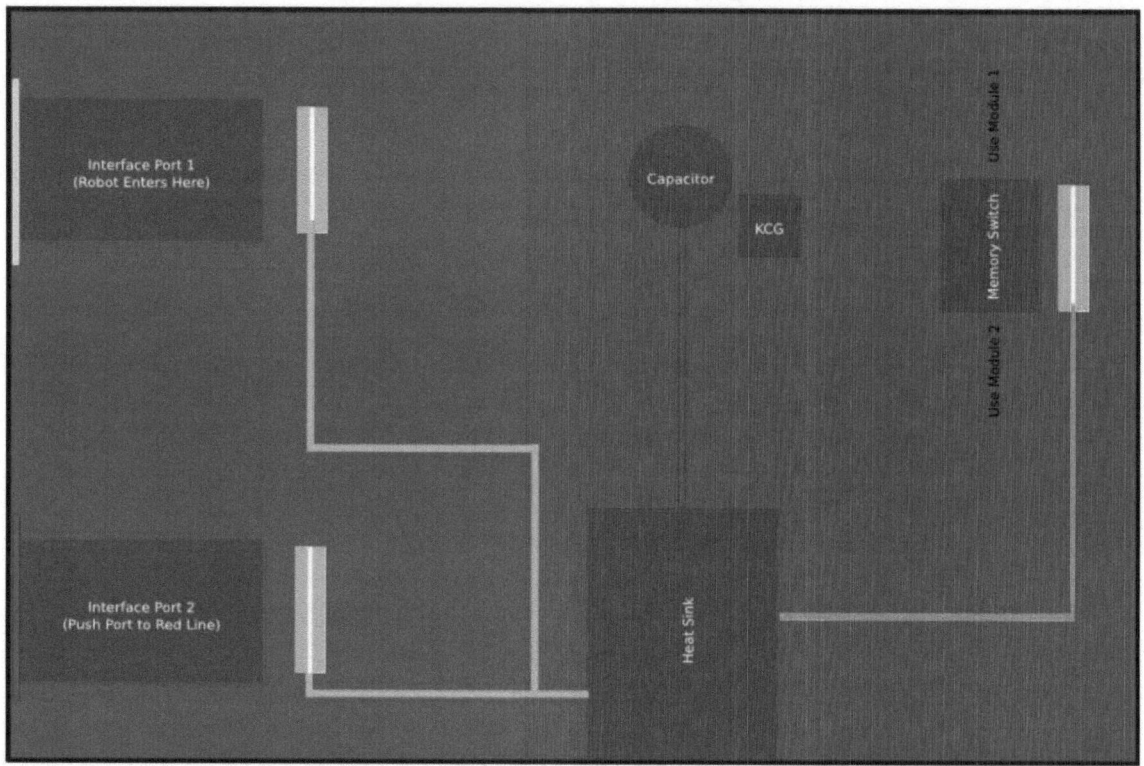

Figure 17-1. The challenge area for the Systems Crash challenge

The CA dimensions are a flat surface area with dimensions of 24×36 inches (2×3 feet) or approximately 61×91 centimeters (refer to Figure 17-1). The greater distance is the CA width.

For purposes of the challenge, the border of the CA must be clearly defined; your robot must enter the challenge area via the Interface Port 1 area designated in Figure 17-1. Once the robot is inside the challenge area, no portion of it may cross the defined boundaries. The only exception is that the robot may leave the CA via Interface Port 1 and return again as many times as is needed within the time limit of the challenge.

Place the heat sink, interface port, memory switch, capacitor, and kinetic charging gears models in their respective locations as defined in Figure 17-1.

Figure 17-2 shows a photograph of the models placed on the printed challenge area mat.

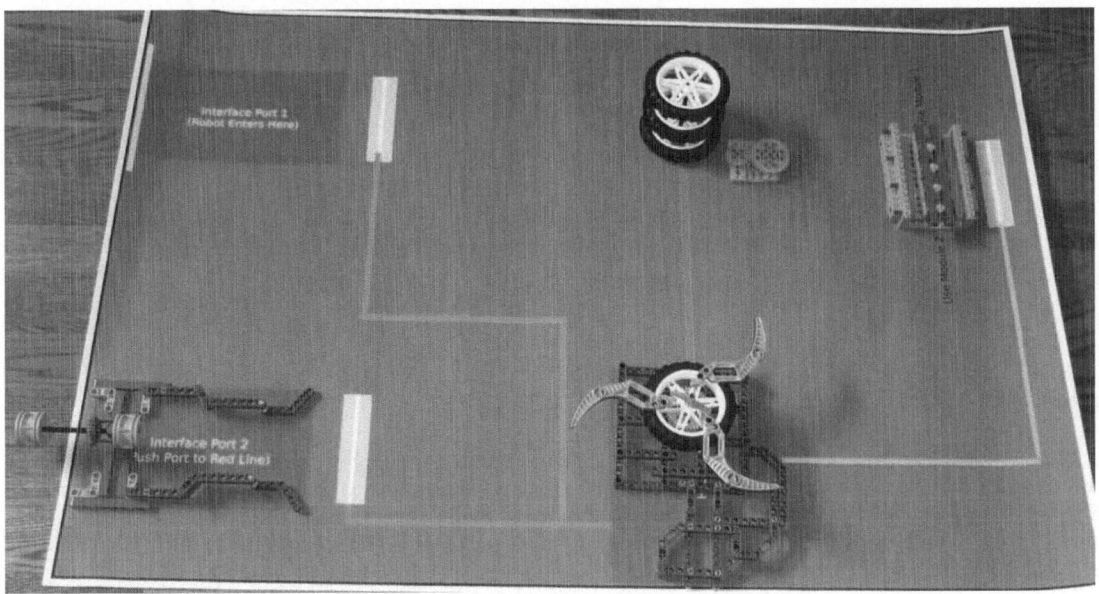

Figure 17-2. The black box's components placed on the mat

Understanding the Challenge Rules

The Systems Crash challenge has a number of rules that must be followed by teams or individuals. If multiple teams are running the challenge, rules can be altered or removed if all players are in agreement. The rules for the Mars Base Alpha Systems Crash Challenge are as follows:

1. The robot created to attempt the missions must be built using parts from a LEGO MINDSTORMS NXT robotics kit (Retail or Education version 1.0 or higher).

2. A minimum of one robotics kit must be used to build the robot.

3. A maximum of two robotics kits may be used to build the robot.

4. If a LEGO MINDSTORMS NXT Education Resource Set is used to build the heat sink and other components, any parts remaining in the Resource Set may also be used in the robot design.

5. Only NXT motors may be used in the robot's design.

6. Approved sensors are the Ultrasonic, Sound, Touch, Color, and Light (NXT versions only).

7. The challenge time limit is 2 minutes. The challenge ends when the time limit expires.

8. The robot may be placed outside the mat/challenge area or sitting on the Interface Port 1 area before the challenge begins. The starting position must be agreed on by all challenge participants.

9. Bonus points are scored based on successfully completing all mission objectives and avoiding secondary obstacles.

10. Tape (any color) may be used and placed anywhere on the challenge area to assist the robot with navigation. Tape may be used to define the CA boundaries as well as to provide lines to follow or points on the CA. There is no limit to the amount of tape that may be used.

11. Tape, Velcro, and other items may not be used to secure the models to challenge area.

Understanding the Mission Objectives

There are five primary mission objectives; points will be awarded for any mission objectives completed, and bonus points will be awarded if all primary missions objectives are completed successfully. Mission objectives may be attempted in any order, but bonus points will be awarded if Interface Port 2 is repaired first; this will allow for better monitoring of the other repairs that are being made. The mission objectives follow:

Mission objective 1—repair Interface Port 2: The robot must push the sliding interface port so that it crosses the boundary of the black box (defined by the red line on the color mat).

Mission objective 2—repair the heat sink: The robot must close the open link to repair the heat sink, allowing heat to flow from the CPU to the walls of the black box.

Mission objective 3—toggle the memory switch: The robot must push the switch in the opposite direction of its starting position. This will represent switching the black box from using Memory Module 1 to using Memory Module 2.

Mission objective 4—avoid the capacitor and kinetic charging gears: The robot must avoid touching both of these components. Penalty points will be awarded if they are disturbed.

Mission objective 5—exit the black box: Before the time limit expires, the microbot must exit the black box via the Interface Port 1 area.

Mission Objective 1: Repairing Interface Port 2

The interface port repair objective will require your robot to push a sliding mechanism in such a way that it crosses the boundary of the black box. Here are the requirements for setting up the interface port:

1. When the challenge starts, the interface port's sliding mechanism must be placed so that it is as far as possible from the boundary or edge of the black box. If you are using the color mat, push the mechanism away from the red line that indicates the boundary. Figure 17-3 shows the initial starting position of the Interface Port 2 mission objective.

2. The microbot must push the sliding mechanism so that the any portion of the sliding mechanism crosses the boundary, as shown in Figure 17-4.

3. No other part of the interface port model must be moved. (Or, challenge participants can agree on an allowable amount of movement if desired.)

4. If the sliding mechanism crosses the boundary but is later (accidentally) pulled back, no points will be awarded. The sliding mechanism must be across the boundary when the time limit expires.

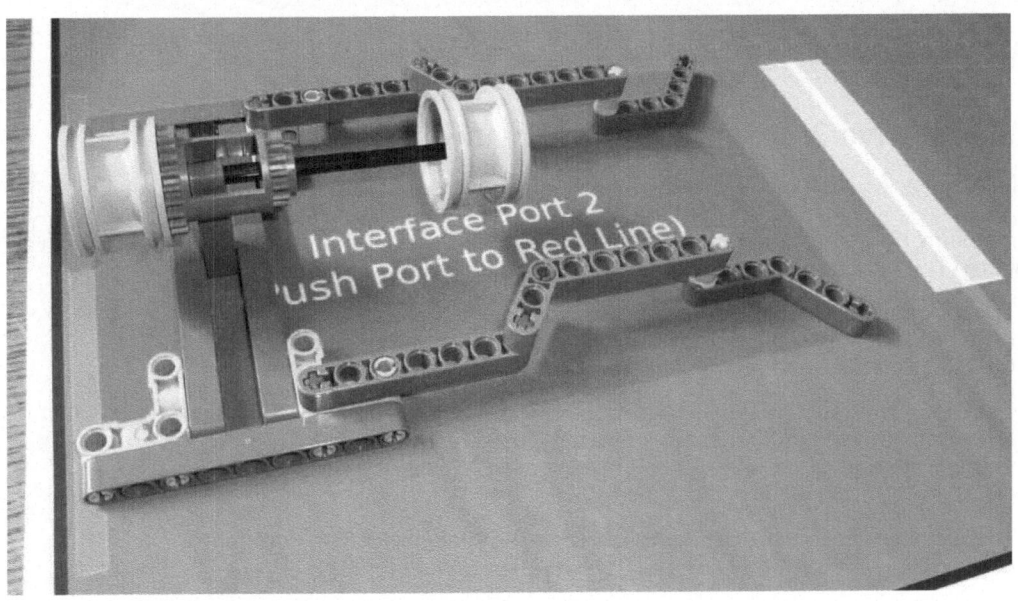

Figure 17-3. The starting position of the interface port

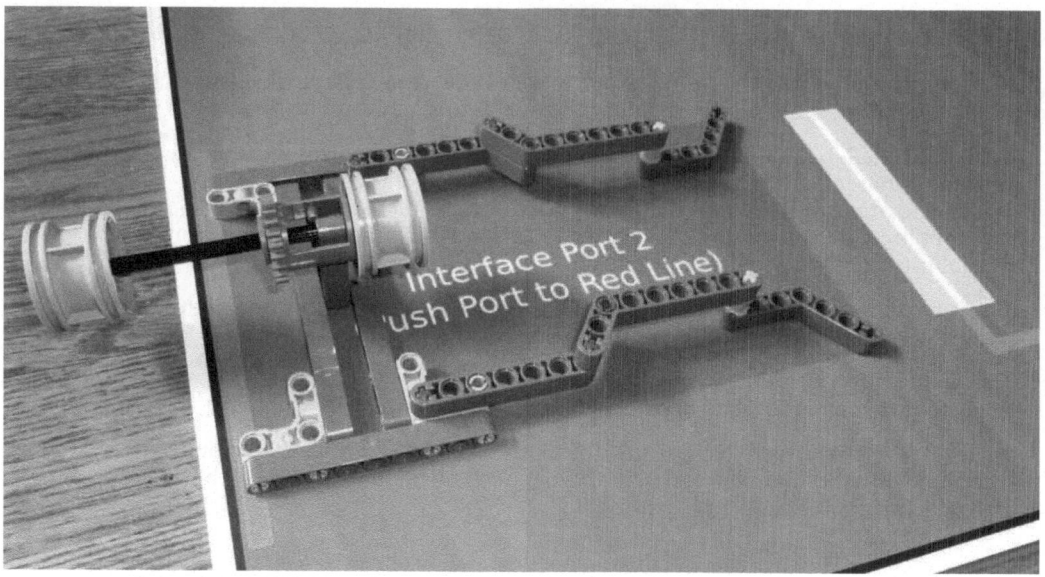

Figure 17-4. The ending position of a repaired interface port

Mission Objective 2: Repairing the Heat Sink

When the challenge begins, the heat sink will not be connected to the wall of the black box. The robot must repair the heat sink by pushing or pulling the swing arm connected to the CPU fan so that it touches its matching piece on the heat sink device. When the challenge starts, position the swing arm as shown in Figure 17-5. Additional rules for the heat sink objective are as follows:

1. The robot must not disturb the fan. Touching the delicate fan sitting on top of the CPU will damage it and may result in the CPU becoming damaged when the black box is turned on; penalty points will be given for touching the fan.

2. The swing arm must be touching its matching arm at the end of the time limit. If the arm is accidentally reopened and not closed before the time runs out, no points will be awarded for this objective. Figure 17-6 shows a successfully closed swing arm and a repaired heat sink.

3. The robot must avoid disturbing the CPU, fan, and heat sink. (Or, competitors may agree on an allowable amount of movement of the model.)

Figure 17-5. The heat sink's starting position

Figure 17-6. Successful completion of the heat sink objective

Mission Objective 3: Toggling the Memory Switch

Memory Module 1 is damaged, but a switch will allow the black box to utilize backup Memory Module 2. The microbot must push the switch so that the rocker arm shown in Figure 17-7 is pointed in the opposite direction of its starting position. Here are some additional rules related to the memory switch objective:

1. When the challenge begins, the rocker arm shown in Figure 17-7 must be pushed all the way in one direction. If you are using the mat, point it in the direction of the Use Module 1 label.

2. When the time limit expires, points will be awarded if the switch has been pushed to the opposite side, as shown in Figure 17-8. If you are using the mat, it must be pointed in the direction of the Use Module 2 label.

3. The robot must be careful to avoid touching either memory module. Penalty points will be given if a memory module is touched by the robot. Competitors may agree on an allowable amount of movement should the robot touch a memory module model.

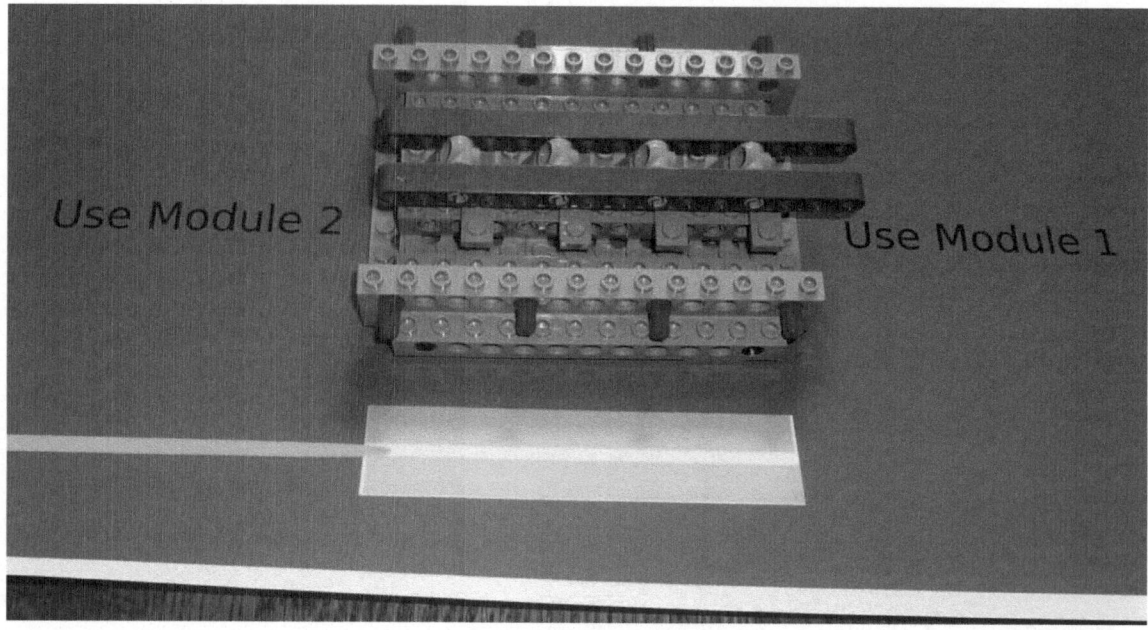

Figure 17-7. Starting positioning for the memory module objective

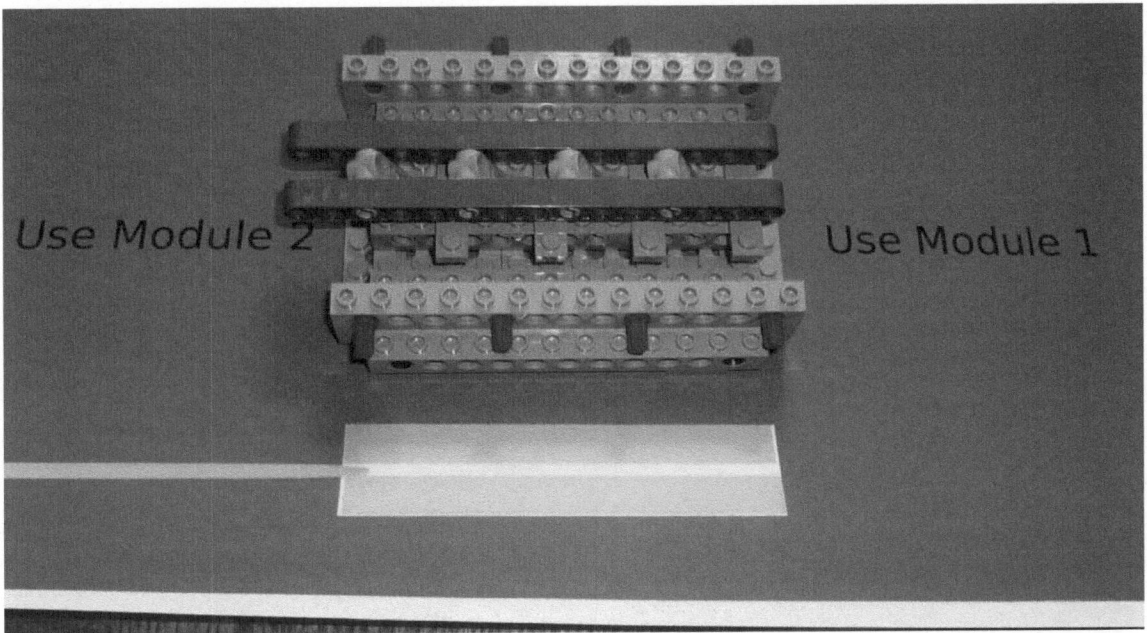

Figure 17-8. Successful completion of the memory module objective

Mission Objective 4: Avoiding the Capacitor and Kinetic Charging Gears

The robot must avoid touching the capacitor and kinetic charging gears. Further rules related to these models are as follows:

1. When the challenge begins, the capacitor and kinetic charging gears models are placed as shown in Figure 17-9.

2. If the kinetic charging gears are touched by the robot, penalty points will be awarded.

3. If the capacitor is touched, the challenge ends immediately, because the microbot will have become damaged from the discharge of electricity. Any mission objectives completed before the challenge ends (due to touching the capacitor) will still be rewarded.

Figure 17-9. Placement of the capacitor and kinetic charging gears

Mission Objective 5: Exiting the Black Box

The robot must exit the black box before the time limit expires. The microbot cannot remain in the black box, and the following rules apply to this objective:

1. When the challenge time limit expires, the microbot must be completely off the matt (or out of the defined challenge area).

2. If any portion of the microbot is sitting on the boundary of the black box, penalty points will be awarded.

Scoring the Challenge

The Challenge Score will be determined using the following system:

- Mission objective 1 completed successfully: 30 points

- Mission objective 2 completed successfully: 20 points

- Mission objective 3 completed successfully: 15 points

- Mission objective 4 completed successfully: 20 points, divided as follows:
  - Capacitor untouched: 10 points
  - Kinetic charging gears untouched: 10 points
- Mission objective 5 completed successfully: 10 points
- Kinetic charging gears touched: –15 points (points only deducted once)
- Capacitor touched: –15 points (points only deducted once)
- Interface Port 2 disturbed: –15 points (points only deducted once)
- Heat Sink disturbed: –15 points (points only deducted once)
- Fan on heat sink touched: –5 points (points only deducted once)
- Memory module disturbed: –10 points (points only deducted once)
- Robot crosses the black box boundary: –5 points per incident
- Incomplete exit of the black box before time expires: –5 points

Earning Bonus Points

If you choose, you may award the following bonus points:

- All five primary mission objectives successful: 25 points
- Microbot enters and exits the black box only once: 10 points

Using the Mars Base Alpha Mission Scoring Form

Use the Mars Base Alpha Scoring Form to track the completions of the mission objectives and to tally the final challenge score.

Note   The Mars Base Alpha Mission Data Form and the Mars Base Alpha Scoring Form may be downloaded at www.marsbasecommand.com by clicking the Mars Base Alpha Scoring button.

Frequently Asked Questions

Following are some frequently asked questions and their answers:

1. *Do I have to successfully complete all the primary mission objectives?*
   No. Points will be awarded for any objectives completed, but the bonus points for all missions completed successfully will not be available. (Thankfully there are no real lives at stake here, right?)

2. *What if my robot completes all the mission objectives but touches the capacitor as it's leaving?*
   You will get all points for Interface Port 2, the heat sink, and the memory module (objectives 1–3), plus those points for avoiding the kinetic charging gears. You will not receive any bonus points, however, as your robot did not leave the challenge area and it failed the "avoid capacitor" portion of mission objective 4.

3. *Is there a height or weight limit for the microbot?*
   No. But you are limited in the parts you can use: one NXT robotics kit plus leftover parts from one Resource Set (if one is used). Remember, the smaller the microbot you build, the better.

4. *What happens if my microbot touches a model but doesn't move it?*
   Touching (but not disturbing) the heat sink and Interface Port 2 models is allowed, but touching the fan (on top of the CPU) or kinetic charging gears results in penalty points. Touching the capacitor incapacitates your robot and ends the challenge.

5. *What if I encounter a situation that's not covered by the challenge rules?*
   Make your own ruling on the matter that is in agreement by all parties participating in the challenge.

Adding Novice and Expert Rules

As an alternative to the standard mission challenge rules, some novice rules have been created that will make the challenge a little easier to successfully complete. These novice rules are as follows:

1. The challenge time limit is doubled to 4 minutes.

2. The capacitor and kinetic charging gears can be removed from the challenge area.

As an alternative to the standard mission challenge rules, the following expert rules have been created to make the challenge a bit more difficult to successfully complete:

1. The challenge time limit is cut in half to 1 minute.

2. The Interface Port 2 repair must be performed first.

3. Touching the capacitor destroys the robot and the black box; no points are awarded.

Summary

The Mars Base Alpha Systems Crash challenge is likely to be considered the most difficult challenge in this book. You have limited time to perform three repairs, all while avoiding touching other components that will result in the microbot's destruction. Add to this the fact that your robot must enter and leave the black box before the time expires, and you have a lot of work ahead of you.

Remember, though, it's not about winning or losing. Your first goal should be to attempt the challenge and obtain a score. You should then attempt to improve your robot, either by modifying its design or its programming, or both. Run the challenge again, and try to improve your score; an increasing score is one of the best methods for determining if your robot building and programming skills are progressing.

Remember, if you don't like the rules, change them! Give yourself more time or allow third-party sensors or whatever will make the experience more enjoyable for you. Make the challenge fun, not tedious.

CHAPTER 18

■ ■ ■

Mars Base Lambda

Job Well Done

Mars Base Alpha, Section D, Commander's Office
September 5, 2062 at 8:15 AM (Greenwich Mean Time)

Commander Evans was standing behind his desk, back to the door, as he stared out the large window at the latest technicians arriving by rover. Another twelve specialists had just passed quarantine and were on their way to orientation in the Rec Hall. Evans looked at this watch and calculated that he still had fifteen minutes until he was scheduled to give the group a welcome speech.

Time to let her know, he thought, pressing the intercom button on his desk. "Lieutenant Raleigh, please report to the Commander's Office."

Evans flexed the fingers on both his hands. The prosthetic was repaired and he could no longer feel the slight tingle in the fingertips. And his eye had finally stopped stinging after the metal fragments had been removed. *Doc Homansky is a miracle worker*, he thought.

"On my way, sir," came Lieutenant Raleigh's response a few seconds later from the speaker on the desk.

The past few weeks had been a nonstop flurry of activity as Alpha had completed repairs on all systems and Beta had become fully upgraded and was now receiving staff transferred from Alpha and Gamma. Lieutenant Raleigh had been in charge for the complete overhaul of Alpha while Evans had been temporarily relieved of duty to recover and for a variety of testing on his vision and the use of his repaired hand. He hated not being available to his second-in-command, but she had done her job and done it well.

Evans opened the door to his office and stepped into the Control Center. "Alright, everybody, inside now. Hurry."

Doctor Kim Homansky followed the commander into the office, along with Mason Gerard and his sister, Ginny. Louie Pitchersky brought up the rear.

Evans closed the door as the four visitors each took a seat against the wall.

"How's the hand," asked the doctor. "Has the tingling sensation gone away?"

Evans smiled. "Completely. Thanks, Doc."

Evans dropped into his chair, rolled forward, and typed a command into the keyboard built into the desktop.

"How's it feel to be back on solid ground?" asked Evans, looking up at Ginny Gerard and Louie Pitchersky.

"I may never want to step onto a lab satellite again," replied Louie.

Ginny nodded. "I think I'll let Mason have the next two or three rotations until I get my nerve back."

"How long were you out there waiting for the pick up?" asked Kim.

"Fourteen hours. We had plenty of food and water, but the high-bandwidth antennae was damaged so we couldn't download any movies to pass the time," said Louie. "We played solitaire until the data pad batteries died, and then we napped."

"You snore," laughed Ginny. "Not loud, though."

Evans laughed. "We could have piped in some music to you over the comm channel."

"Now you tell us," said Louie.

The conversation was interrupted by a knock on the door.

"Game faces, folks," whispered Evans. "Come in!"

The door opened and Lieutenant Raleigh walked in, her hair pulled back in a ponytail. Her hands were dirty, and she had a smear of oil on her left cheek. She stopped almost immediately at the sight of the other four occupants of the office.

"I'm sorry, sir," she said. "I've been helping Standridge replace the transmission in the Number Five rover."

Evans waved the comment off. "Is it fixed?"

"It's now the fastest rover in the garage, sir. Should get you to Lambda in less than twelve hours."

"Awesome," replied Evans. "I'll, uh, want to take it out for a test drive later. You know, make certain everything is in working order."

"Of course, sir," Kristie replied with a smile. "I'll follow you in the Number Three rover. Just in case. Murray's Rock and back ought to be a good test."

Kim shook her head. "I don't want to hear this, folks. Racing rovers isn't on my list of approved staff activities."

Evans nodded. "You're right, Doc. Let's schedule the rover test later, Lieutenant."

"Yes, sir," replied Kristie. "You needed to see me, sir?"

Evans nodded to the chair in front of his desk. "Have a seat."

Kristie looked over at Kim and the other visitors, a puzzled look on her face. She sat in the offered chair and looked directly at the commander.

Commander Evans leaned forward, folding his hands in front of him on the desk. "I've got a big problem, Lieutenant." He was no longer smiling.

Kristie maintained eye contact with the commander, but she inhaled slowly. She nodded. "OK," she said. "What's happened, sir?"

Evans jabbed his finger at the screen on his desk. His jaw clenched, and he shook his head. "I just received a priority message from Command, telling me that they're transferring your replacement over from Gamma in fifteen days."

Kristie gasped. "I don't understand, sir. What did I do? You know I would never…"

"I don't want to hear it, Lieutenant," said Evans. "It's a done deal. Lieutenant Baker has already received his orders."

"But sir," said Kristie. "Shouldn't there be some sort of hearing? I don't even know what I've done?"

"You don't know what you've done, Lieutenant?" asked the commander. "Lieutenant, what you've done…"

Kristie blinked, waiting to hear the complaint against her.

Commander Evans leaned back in his chair, a smile growing on his face. "What you've done, Kristie, is an outstanding job of keeping this base operational. You're responsible for saving numerous pieces of expensive equipment, not to mention my hand. And you got two of your fellow specialists back to Mars in one piece."

Kristie was speechless. She turned to stare at the four smiling individuals sitting against the wall.

"Job well done, Kristie," said the commander, standing up and holding his hand out to Kristie.

Kristie shook the commander's hand. "I, uh…thank you, sir. Thank you."

"I wasn't able to get the entire base staff here today; there's still a lot of jobs that need to be done," said Evans.

"Of course, sir."

"But I wanted these four here for a very special reason. You did a great job while I was out of commission, and Mars Base Command feels the same way. But I wasn't kidding about Baker coming here in two weeks to take your job."

Kristie's smile disappeared. "Sir?"

Commander Evans smiled.

Keep a Clear Head

Mars Base Lambda, Section D, Command Center
September 19, 2062 at 6:22 AM (Greenwich Mean Time)

"Welcome to Mars Base Lambda, Major Raleigh. I'm sorry. Commander Raleigh."

"Thank you, Lieutenant Gilbride. I'm still adjusting to the promotion, so I'll answer to either," replied Kristie.

"Yes, ma'am," said Gilbride.

"It looks like we've got a lot of work, Lieutenant," said Kristie.

"I thought you might like to ease into the job, Commander. I've got a list of all the upgrades and build-outs that we'll be experiencing over the next six months, but we can go over it later if you'd like some time to settle in and get unpacked."

"Not my style. Lambda will be the largest of the bases when we're done. Mars Base Command has informed me that the next batch of specialists from Earth is due to arrive in three months' time. Let's make certain they have a place to work, sleep, and eat. Let me grab a cup of coffee, and we'll go over that list in my office in five minutes."

Gilbride smiled. "I'll send it to your desk and be with you shortly."

Kristie walked into her new office. Three sealed containers full of her personal belongings were sitting against the far wall, but she ignored them and took a seat. She logged into the system with her new credentials and was surprised to find a video message waiting from Commander Evans. She pressed the Play button.

"Kristie, I know you're busy, so I won't take a lot of your time. I just wanted to wish you luck with Lambda. If you need some extra staff in the coming months, don't be afraid to ask. I'll see what I can do to help you get Lambda fully operational in five months. Keep a clear head, and remember that you kept Alpha together when it needed you the most. You were a commander well before you got the promotion. Speak to you soon."

The video ended. Kristie smiled as she saved the file and opened the list that Lieutenant Gilbride had forwarded to her. The young lieutenant knocked on her door, and she motioned him to take a seat.

"It looks like this list is prioritized based on current staff availability and heaviest load on power resources."

Gilbride nodded. "Yes, ma'am." He sat straight, hands on his lap.

"Nice work, Lieutenant."

Kristie looked at the list, then at Gilbride. "How about we head down to the mess hall and grab something to eat while we go over this list? Do you mind if I call you, Chris, Lieutenant?"

"Uh, no ma'am. That's fine."

"If you're going to be my second-in-command, I think we should get to know one another. You can start by telling me about the runaway maintenance bot I heard stories about."

Gilbride's eyes widened. "You heard about that?"

"It's a commander's job to know everything about her base. And it's your job to be able to step in and take over my duties if necessary. I need to trust you, and you need to trust me. So let's go grab some food and talk about Lambda. Sound good?"

Chris nodded. "Deal. And I'd like to hear more about the meteor strike on Alpha if you don't mind."

Kristie stood and walked with the lieutenant out of the office. The Command Center staff continued with their duties without looking up from their screens.

"Sure, I'll tell you all about it," said Kristie as she followed Chris down the hallway. "By the way, how's your knowledge of robotics?"

"It's been a hobby of mine since I was a kid," said Chris. "My mom and dad bought me the LEGO MINDSTORMS 2038 kit when I was ten, and I've dabbled all my life. Why?"

Kristie smiled. "You just never know when a bit of building or programming experience might come in handy."

CHAPTER 19

■ ■ ■

Designing Your Own Challenges

Since writing *LEGO Mindstorms NXT: The Mayan Adventure* back in 2006, I've been asked many times by teachers, parents, and students how I go about designing a challenge. They ask questions like, "Do you start with the story?" or "Do you build the models first?" I've been asked if I always know how the story ends, and I even had one student request to be a character in a future story.

I'll try to answer most of these questions in this final chapter, but I would like to state that there is no cookie-cutter formula. With *The Mayan Adventure*, I had most of the story worked out in my head long before I ever started designing the challenges the characters would encounter. But the exact opposite happened with the sequel, *The King's Treasure*. For that book, I had most of the challenges thought out and about half of the challenge areas built or at least prototyped before I ever started writing the story that would use those challenges.

With this *Mars Base Command* book, the process was a bit mixed. For the first two challenges, I already had the stories plotted out and knew what the goals of the challenges would be. For the last two challenges, I developed the challenges first and then tailored the story so the characters would encounter these challenges in a fictional setting.

For the rest of this chapter, I'll introduce you to some of the tools and methods I use for designing challenges. I don't have a set of rules that I adhere to, and I don't always know how a challenge will actually work until I build and test it. But if you would like to create your own challenge for your classroom or maybe for a friend or robotics team to tackle, maybe you'll find something helpful here to get you started.

Writing the Story

A story isn't an absolute requirement for your challenge, but it does help provide an incentive for successfully completing it. The story can do many things. First, it can be used provide the robot builder with consequences of failure. The story tells the builder what has happened and why a robot is needed. If a human can solve the problem without the use of a robot, there's no robot challenge. The story has to provide a good reason for taking the human element out of the equation and putting in a robot to solve whatever problem the class, team, or even Mars base is facing.

The story doesn't have to involve life or death choices, either. You can easily create a story that simply substitutes a robot wherever a human might need to be involved. You can create a science fiction story, a scary story, or even a fairy tale—the story simply gives the robot builders a setting for their robots to solve a problem.

When creating a story, try to keep the characters to a minimum. There's no need to have more than five or six characters to move the story along, and the fewer you have, the easier writing the story will be. For example, I encountered an issue with the "Systems Crash" storyline in Chapter 14, because it had a large number of characters that became quite confusing during the first draft of the story. Keep in mind that you're goal with a story is to provide background or incentive (or both) to the reader to attempt to solve the challenge you are creating.

Creating the Challenge

There are two ways, again, that a challenge can be created. The first occurs when you have an idea for the challenge and create the story first. I've used this method before, because sometimes, writing the story helps me to develop (in my mind) the vision I have for the final challenge. Other times, I have an idea for the layout of the challenge I want to create, and then I write the story later and make certain it has elements mentioned in it that match up to the actual challenge the players will attempt.

Whichever method you use, you'll find that designing the challenge is, well, challenging. You've got to designate the size of the challenge area and determine how many models will be created as obstacles to the robot's success. You also have to prototype those models, test them out with your own robot design, and the refine them as you discover things that don't quite work.

For example, in the "Plan B" storyline, my first attempt at designing the solar collector was too wobbly and had a tendency to fall apart if the robot ran into it with any amount of speed. After a few tweaks here and there, I was able to design a solar collector that didn't fall apart and was actually a bit easier for my test robot to rotate.

With most of the challenges I design, the challenge area is my first consideration. I ask myself, "How much space do I want to allow the players and their robots to roam?" With almost all of the Mars Base Command challenges, I've limited the challenge area to a 2-foot × 3-foot space for the mat. In some instances, there are penalties if the robot leaves the challenge area, and in other instances, the robot can roam wherever it wishes without penalty. In these latter cases, the mat simply serves as a base for placing the models in a way that gives the robot something to do.

Speaking of the mat, I've been asked questions about how I create these. I don't have the room in this chapter to create a complete tutorial on the process, but I can give you the basics. Here we go:

1. Initially, I test my mat design with my model prototypes using simple poster board as the design surface. Once I have the basic layout, I use the open source application (it's free) called Inkscape to create my mats in color.

2. I typically create three layers when designing a graphic of the mat:

    • Background: The color used for the majority of the surface)

    • Models: For defining boxes where models will be placed as well as text hints on the mat indicating those locations

    • Extras: For adding extra graphical elements that aren't necessarily required but give the mat a cool look

3. After creating the Background layer, I select the color (or texture, such as the Martian surface used in the Storm Front challenge) and fill in the 2-foot × 3-foot surface. I then lock that layer so it cannot be altered.

4. I next create the Models layer, and based on my testing with poster board and my model prototypes, I create small rectangles or squares to indicate on the mat where to place the models. I then lock this layer so it can't be changed accidentally.

5. Finally, I create the Extras layer, where I add additional elements such as the warning labels in the Plan B challenge mat or the electrical trace lines in the Systems Crash mat.

6. After the mat is created, I save it as a PDF file and take it to my local print shop to print a copy in black and white (it's cheaper than color) for a final test.

7. If the black and white mat with models added works with my test robot, I print out a full color mat (which is a bit more expensive) for actual challenge runs.

Inkscape has plenty of online tutorials (many are videos) that show you how to properly use the application and its various menus and toolbars. You can download the free Inkscape application by visiting www.inkscape.org.

As I'm creating the mat, I'm also refining the prototype models that the robot will interact with by turning, moving, lifting, and other operations. The mat and the models go hand-in-hand, and it's difficult to design one without the other. You may find, as I have many times, that a model changes in shape or size during testing, so don't get too attached to a specific layout on the mat. If necessary, eliminate a model that just won't work, or try moving the existing models around to allow enough space for a robot to move.

Ultimately, your goal is to create both a challenge area and a number of models that will make sense to the challenge participants and not be overly difficult. An illogical layout (or an extremely difficult layout with too many models in a space that's too tight) will frustrate players and result in a challenge that just isn't fun. That's why it's important to test your challenge and models with your own prototype robot and maybe ask a few friends or other teams to give you feedback on your challenge design.

Finally, keep in mind that a challenge does not require an actual challenge mat. You can use a variety of resources such as tape, poster board, and paints to create a challenge area that doesn't require printing an actual mat. The goal is a fun challenge, not to see who can create the craziest mat (but that's fun, too, if you have the time and funds).

Designing the Models

The models are (to me) the most frustrating part of designing a challenge. In this book, I was limited to a single Resource Set for all the parts to create the models in each challenge. While you're certainly not limited in this manner, do keep in mind that not everyone will have the same LEGO parts that you may have, so in many instances, you may want to consider designing your challenge to use models that consist of common items: balls, boxes, paper clips, and more. There's nothing to prevent you from using a pen in a challenge and calling it a steel reinforcement beam that must be lifted by a robot to repair a damaged building.

If you decide to use LEGO components to build your models, you'll want to keep two things in mind as you create your prototypes:

- Keep it simple.

- Do not get attached to a particular design.

You've got to be flexible here, so don't get stuck in a design by telling yourself that you absolutely must use part X in the model. The models are supposed to serve as stand-ins—simulations—for real world objects. You want your models to be as easy to build as possible while also making certain they're not so flimsy that they fall apart when a robot bumps them.

Design your first model, run a few tests with your test robot, and see what works—and what doesn't work. Is a part too tight? Is a rotating arm too difficult to turn? Is a model too short? Too tall? Are little pieces your robot must pick up so small they can't be easily grasped or scooped up? Make changes based on the problems you encounter, and never be afraid to completely dismantle a model that just won't work properly.

Always look at your models with the goal of reducing the part count. Because I am using the Resource Set, I have a limited number of each and every part. If I can eliminate the use of two parts X and one part Y from the first model, this leaves me with more of parts X and Y for the next model. I cannot tell you how many times I ran out of a particular connector piece or certain length of beam and had to redesign an earlier model to free up one or more pieces that another model absolutely required.

Models are tricky, so be sure to give yourself plenty of time to prototype, test, and refine yours. Because the models are typically the key to a fun challenge, you want to make certain that the models you do build and include in your challenge work properly, are easy to duplicate by your challenge participants, and make sense in relation to the story and the challenge area you've created.

Setting the Rules

A good challenge must have rules. But you don't want to have so many rules that the challenge is no longer fun. You'll want to start with the basics: Define the challenge area so that players know where their robots can move and where they cannot. Define a time limit, but don't make it too short; give the players and their robots just a bit more time than you think necessary (you'll often be able to determine a time limit based on your own testing with your test robot).

Points are useful for various tasks. Make certain that every robot gets at least some points for just making an attempt at a challenge. This is especially important for younger competitors who need to be encouraged to refine their robots and find new ways to solve challenges. Give plenty of bonus points, too, and not just for the difficult models. Try to look at your challenge and see where players will encounter problems and then award more points for creative solutions or even fast finishes.

Players are always going to ask for more information on rules, so my overall rule has always been that it's OK to change the rules when all players are in agreement. Rules should not overshadow the fun aspect of building and programming a robot to solve a challenge. When in doubt, throw out rules if the challenge is too difficult, or add new rules if teams are scoring so many points that it's hard to distinguish a winner.

Having Fun

I love designing challenges. Some are difficult; some are easy. When I set out to create a new challenge, I'm not looking to frustrate the players. I'm not trying to overwhelm them with rules and limitations. What I'm trying to do is put a set of obstacles in front of their robots that the players can observe, measure, test, play with, and talk about—all before they ever start building a robot.

I've always enjoyed a challenge. It gives me a chance to push my knowledge, learn something new, ask others for advice, and, hopefully, have some fun. I hope that my Mars Base Command challenges have inspired you to think like a challenge designer. I also hope you have an idea in your head for your own challenge, something that's never been seen before and that your friends are sure to enjoy.

Don't look at challenge design as hard work; treat it as play. Find the right mix of challenge rules, model design, and story development to give players a good reason to set aside some time and accept your challenge.

Have fun!

Index